T0135510

RATIONAL TUNNELLING, 2ND SUMMERSCHOOL, INNSBRUCK, 2005

ADVANCES IN GEOTECHNICAL ENGINEERING AND TUNNELLING

13

General editor:

D. Kolymbas

University of Innsbruck, Institute of Geotechnical and Tunnel Engineering

In the same series (A.A.BALKEMA):

1. D. Kolymbas (2000), *Introduction to hypoplasticity*, 104 pages, ISBN 90 5809 306 9

2. W. Fellin (2000), *Rtteldruckverdichtung als plastodynamisches Problem, (Deep vibration compaction as a plastodynamic problem)*, 344 pages, ISBN 90 5809 315 8

3. D. Kolymbas & W. Fellin (2000), *Compaction of soils, granulates and pow-ders - International workshop on compaction of soils, granulates, powders*, Innsbruck, 28-29 February 2000, 344 pages, ISBN 90 5809 318 2

In the same series (LOGOS):

4. C. Bliem (2001), *3D Finite Element Berechnungen im Tunnelbau, (3D finite element calculations in tunnelling)*, 220 pages, ISBN 3-89722-750-9

5. D. Kolymbas (General editor), *Tunnelling Mechanics, Eurosummer-school, Innsbruck, 2001*, 403 pages, ISBN 3-89722-873-4

6. M. Fiedler (2001), *Nichtlineare Berechnung von Plattenfundamenten (Nonlinear Analysis of Mat Foundations)*, 163 pages, ISBN 3-8325-0031-6

7. W. Fellin (2003), *Geotechnik - Lernen mit Beispielen*, 230 pages, ISBN 3-8325- 0147-9

8. D. Kolymbas, ed. (2003), *Rational Tunnelling, Summerschool, Innsbruck 2003*, 428 pages, ISBN 3-8325-0350-1

9. D. Kolymbas, ed. (2004), *Fractals in Geotechnical Engineering, Exploratory Workshop, Innsbruck, 2003*, 174 pages, ISBN 3-8325-0583-0

10. P. Tanseng (2005), *Implementations of Hypoplasticity and Simulations of Geotechnical Problems*, in print.

11. A. Laudahn (2005), *An Approach to 1g Modelling in Geotechnical Engineering with Soiltron*, in print.

12. L. Prinz von Baden (2005), *Alpine Bauweisen und Gefahrenmanagement, (Alpine construction methods and risk management)*, 234 pages, in print.

ACKNOWLEDGEMENT

The financial support of the Summerschool by the following organisation is gratefully acknowledged:

Federal Ministry for Education, Science and Culture
Minoritenplatz 5
A 1014 Vienna
Austria

RATIONAL TUNNELLING 2nd SUMMERSCHOOL, INNSBRUCK, 2005

Edited by

Dimitrios Kolymbas & Andreas Laudahn

University of Innsbruck, Institute of Geotechnical and Tunnel Engineering

The first three volumes have been published by Balkema
and can be ordered from:

A.A. Balkema Publishers
P.O.Box 1675
NL-3000 BR Rotterdam
e-mail: orders@swets.nl
website: www.balkema.nl

Titelbild:
Die Luegbrücke, Brennerautobahn
Steiner, Private Photosammlung Innsbruck/Tirol

Bibliographic information published by Die Deutsche Bibliothek

Die Deutsche Bibliothek lists this publication in the Deutsche National-
bibliografie; detailed bibliographic data is available in the Internet at
http://dnb.ddb.de.

ISBN 3-8325-1012-5

ISSN 1566-6182

Logos Verlag Berlin
Comeniushof, Gubener Str. 47,
10243 Berlin
Tel.: +49 030 42 85 10 90
Fax: +49 030 42 85 10 92
INTERNET: http://www.logos-verlag.de

Contents

Blasting methods in tunneling   71
Mark Ganster

Rock reinforcing   199
Dimitrios Kolymbas

Earthquake resistant design of tunnels   269
Christos Vrettos

Safety in tunneling

Yannis Bakoyannis

Greek Ministry of Public Works, Special Service for Road Tunnels and Underground Works

Abstract: Safety in tunnelling has been inherently related to risks. As long as zero risk does not exist, even in more single everyday conventional activities, absolute safety should be considered as an illusion. Risk management should be the most effective tool in accomplishment of a safe tunnelling environment. In this context well established risk acceptance criteria and formalized risk management procedures could be of high importance. Design life of the tunnel defines the time domain for evaluation of safety, while performance requirements set the framework of the project where safety has to be fulfilled.

1 Introduction

Safety in tunneling is an interminable issue, which has ever been the main concern of the tunneling community and other involved parties. Generally, tunnels are part of major infrastructure projects and their possible damages may have multilevel consequences.

The first step in rational mastering of safety is an acceptable definition of safety. In most modern codes and regulations the following definition has been adopted: *"Safety is the freedom of unacceptable risks"*

From the above definition linkage of safety and risks is obvious. Nevertheless the crucial term is the word "unacceptable". Historical, public, political, economical and ethical conditions and preferences of individuals and society should be consulted in rationally defining unacceptable risks.

The main advantage of the above definition is the reduction (through elimination) of misconceptions. But are all risks related to safety? Adhering on the safety definition, the answer is yes. Traditionally safety is referred to injuries to persons (fatalities, major and/or minor injuries) and harm to the environment (natural environment and third party properties). In my opinion also other types of hazard consequences (like economic loss or delays or loss of reputation) must be considered for safety evaluation. Though these consequences may stand in someone else's responsibility, no one can ignore side effects, which can produce new risks.

Risk is related with uncertainty. Tunnelling involves many scientific disciplines; each adduces its own point of view in safety and risk thinking, varying in particular focusing and used methodologies and techniques. Synthesis of all these aspects as an interdisciplinary activity provides an integral and overall perception of safety and risk. Safety assumes consideration and actions for the entire life of the tunnel, from the conceptual phase to the detailed design, auditing, acceptance of the tender, construction, insurance, reinsurance, commissioning, operation and decommissioning.

Risk and safety perception have to do with expert and scientific knowledge. Nowadays scientific knowledge provides methods and models that fulfil three basic required characteristics: they should be "sufficiently" safe, simple and true (i.e. to have physical meaning). The critical character of the concept "sufficiently" it is easily perceivable. Considering that the term "sufficiently" cannot be judged rationally on its own, the predominant objective is to optimize the unavoidable compromise between safety-simplicity-truth.

The fundamental question "how safe is safe enough" (sufficiently safe) can be answered in the light of the above definition's concept of unacceptable risk. Nowadays we have to bear in mind the alteration of the traditional industrial societies to the so-called risk societies [1]. Uncertainty and its non-calculable consequences becomes part of our everyday life. Cosmopolitism and universality became the predominant properties of the risk societies. Under such a consideration, at least in areas with societal cohesion (or ambition for societal cohesion) like European Union, adopting a common strategy for risk and safety management becomes on and on an exigency. Moreover, regarding a macro-social scale consideration of safety, overall safety management has to be taken into account to avoid irrational criticism.

Dealing with the concept "sufficiently simple" we have to remind that "things have to be as simple as possible but not simpler". The characterization as simple or simpler nowadays can be determined from the disposed level of computational and calculating power. Many of the simplifications and the assumptions of the rational approaches have been extensively determined by the computational and calculating tools that were provided in the scientific and technological environment of their development. Today the actual acquired levels of computational and calculating power, combined to the scientific knowledge, enforce and necessitate the use of methods and models of greater accuracy and complexity. In such conditions we have to agree what the Occam's razor has to cut and set aside, and what the accepted simplifications have to be, in the actual scientific and technological environment. On the other hand the euphoria and the conceit of technological advances and triumphs mustn't forget the Aristotelian byword: "It is the mark of an educated man to look for precision in each class of things just so far as the nature of the subject admits".

Following the tragic accidents in Mont Blanc, Tauern and Kitzsteinhorn Kaprun tunnels, safety issues have become a matter of debate and gained the concern of public opinion, governments and European Union. These tragic European accidents acted as boosters in accelerating research for improving safety in European tunnels. New generation regulations and directives have been developed and promulgated. Lessons learned for the operation of road and rail tunnels have released actions and safety objectives to prevent hazards, reduce possible consequences, establish organizational and technical requirements, and provide with written operation concepts. But approaches to holistic safety and risk management within the life cycle of a project may be the most important contribution.

Another important learned lesson may be that the concept "it is not possible to happen to us" is only self-contented bighead and nescience. How much certainty more has the negative answer to the next question? Is it possible that an accident like Salang tunnel accident will happen to us? (Note: North of Kabul, Afghanistan, in 1982, in the 2.6 km long Salang tunnel, a military vehicle (carrying explosives?) collided with a tanker truck causing a huge explosion resulting in officially 700 deaths (unofficially the number of victims run into 1500-2000)).

2 Definitions

Implementing of tunneling projects involve people from many disciplines and fields, so it is necessary to dispose a common language and technical terms that should be used in a same sense and similar meaning.

1. **Performance** of a structure is the simultaneous fulfillment of the following requirements:

   - Safety, which is the ability to accept the imposed actions and fatigue strength, without endangering hygiene, health and environment.
   - Serviceability, which is the gratification of the purpose of the function of the structure i.e. limitation of deformations, watertightness, limitation of vibrations etc.
   - Aesthetic appearance, which is the avoidance of evident cracking, maintenance of the geometric features and color etc.

2. **Design life** corresponds to "economically reasonable working life". Working life is the period of time during which the performance of the

tunnel will be maintained at a level compatible with the fulfillment of the performance requirements with accepted low maintenance.

Economically reasonable working life presumes that all relevant aspects are taken into account, such as: costs of design, construction and use; costs arising from hindrance of use; risks and consequences of failure of the works during its working life and costs of insurance covering these risks; planned partial renewal; costs of inspections, maintenance, care and repair; costs of operation and administration; disposal; decommissioning; environmental aspects.

3. **Damage** is the loss of performance within the time.

4. **Hazard** is a possible event or situation or condition which has the potential to have an adverse consequence to human life or health, environment, properties, economy and/or project completion, or some combination of these.

5. **Near miss or incident** is an undesired event that, under slightly different conditions or circumstances could have caused adverse consequence to human life or health, environment, properties, economy and/or project completion, or some combination of these.

6. **Risk R** in general can be defined as an action that jeopardizes something of value. The risk (of an activity with only one hazard with potential consequences C) may be calculated as a combination (or a function) of the probability P that this hazard will occur, with the consequences given the hazard occurs.

7. **Risk acceptance criteria** are the qualitative or quantitative expressions defining the maximum risk level that is acceptable or tolerable for a given tunnel.

8. **Residual risks** are those risks, which are not avoided, eliminated or transferred in the risk mitigation strategy.

9. **Secondary risks** are risks, which arise from actions taken to mitigate other risks or from extensions the original scope of the project. Secondary risks can sometimes be important and always need to be analyzed in their own right.

10. **Individual risk** is the frequency at which an individual may be expected to sustain a given level of harm from realization of specified hazards.

11. **Social risk** is the frequency with which a specified number of people in a given population, or population as a whole, sustain a specified level of harm from the realization of specified hazards.

12. **Risk analysis** is a structured process that identifies both the probability and extent of adverse consequences arising from a given activity, including hazard identification and description of risks, in qualitative or quantitative way.

13. **Risk evaluation** is the comparison of the products of a risk analysis with risk acceptance criteria or other decision criteria.

14. **Risk assessment** includes risk analysis and risk evaluation.

15. **Risk elimination** is the actions intended to prevent risk occurrence.

16. **Risk mitigation** is the actions that reduce risks by reducing probability of occurrence or consequences.

17. **Reliability** is the probability of non-failure (or other adverse performance) of a component of a system or a design: $P_f = P_r\{R \leq L\}$, where P_f is the probability of failure, $P_r\{.\}$ is probability, R is resistance or capacity and L is load or driving force. The variables R and L are aggregate functions of a set of basic variables, which mirror the fundamental uncertainties. Thus reliability is defined as the probability of the complement of the adverse event $P_s = 1.00 - P_f$.

18. **Generalized reliability index** is a standard reliability measure and defined as: $\beta = -\Phi^{-1}-1(1 - P_f)$, where P_f is the probability of failure, and $\Phi^{-1}(.)$ is the inverse Gaussian distribution.

19. **Probability of failure** P_f of an event is the probability that the limit state criterion or failure function defining the event will be exceeded in a specific reference period.

20. **Uncertainty** is a multiple entity that can be primarily classified into two categories: aleatory uncertainties (those that are a function of the physical uncertainty or randomness) and epistemic uncertainties (those that are a function of understanding or knowledge). Sources of uncertainty in tunneling are:

    - The inherent natural, spatial and temporal variability and variation of geomaterial properties and man-made material properties.
    - Random and systematic errors in data collection and testing.
    - Model uncertainty.

- Defaults, omissions and blunders.
- Future legal, social, and economic regime.
- Human behavior.

The most above uncertainties are coupled.

21. **Safety** is absence of unacceptable levels of risk.

3   Tunnel design life and performance requirements

3.1   Tunnel design life

Safety has to be defined in a time context, which is the design life of the tunnel. Design life of tunnels fluctuates from one or two years to some thousands of years. Temporary worksite road tunnels may have a short design life, mining tunnels usually should be designed for 5-30 years, but deep repositories should be designed for thousands of years. For conventional modern tunnels in the domain of civil engineer, design life ranges between 75 and 120 years.

The role of the design life in the design procedure shall be dual: It is a datum for the design and is also related to the other requirements on the designs. Choice of design life affects the characteristics of any phenomenon having a chance to occurre within that time period, namely affects the quantification of the actions. Earthquakes, rainfall and/or flood and hydraulic loads, time depended phenomena (e.g. creep), time depended social, traffic, political, financial data and conditions should be typical examples. On the other hand, from a rational point of view, considering a definite numerical value of design life, de facto will raise the need for demonstration of durability. The conventional design approach with respect to durability, as included into the design codes and guidelines, is largely empirical, based on dcemed-to satisfy rules of prescriptive character. These prescriptive rules (maximum water/binder ratio, minimum binder content, minimum cover, crack width limitation) are related to environmental conditions and completed with other rules such as concerning curing or air entrainment etc. If these rules are satisfied the structure will have an acceptably long but unspecified life.

In modern infrastructure projects the most widespread methods for service life design are:

- Probabilistic methods, investigating service life under exact conditions envisaged.

- Factorial method, as proposed in ISO 15686-1:2000 applying seven factors to the basic value of reference service life, catering for the individual quality, exposure and conditions during use of the building component considered.

- "Engineering design methods for service life design", as proposed in the subtask group of CIB/RILEM Working Commission 175 "Service Life Methodologies", fits in between the two above methods.

The first one probabilistic method and especially the service life design method Duracrete, which enables performance and reliability based calculations, has been recommended for use in tunnels. The method has been recommended for durability design in new tunnels [7, 8], for durability assessment of existing tunnels and for assessment of the consequences of construction defects (e.g. low concrete cover) on the service life and the probability of failure [9].

Gehlen & Schiessl (1999) [7] and Breitenbuchner et al. (1999) [8] presented the probability based durability design for Western Scheldt tunnel where the governing factor being the concrete cover, against the main degradation process that was the chloride attack. Inputs were in the form of stochastic variables. For each event limit states are applied, which are differentiated to Serviceability Limit States (e.g. onset of corrosion, onset of spalling) and Ultimate Limit States (e.g. collapse due to overloading or excessive degradation). The reliability index was chosen for onset of corrosion (SLS) 1.5-1.8, for onset of spalling and corresponding failure of watertightness (SLS) 2.0-3.0 and for collapse of the structure 3.6-3.8.

For ordinary applications of such probabilistic methods, there are a lot of things that need to be done. We need skilled designers, who do have an access to the method and the related techniques. And of course incorporation of the method into manageable design codes should be considered as requisite.

For the design life concept, the end of design life is a very important issue. In general the end of the design life can be defined as the point in time when the foreseen tunnel performance requirements are no longer fulfilled. At the end of design life all possible consequences have to be considered by the designer. In some cases, overstepping of the limit state that governs consequences on surface structures seems to be unavoidable. Especially in the case of aboveground structures of special interest, ensuring rationally inexistence of non-acceptable consequences after the end of design life may be troublesomely achievable.

Another interesting issue is the target design life for the components or subsystems of a tunnel. Some of the existing standards provide the basis for such discussion.

The European Organization for Technical Approvals (EOTA) has developed a table for the assumed working lives of works and construction products with quantitative values. The table is included in the Guidance Paper published by EU in 1999 and presented in Table 1. [3]

Assumed working life of works (years)		Working life of construction products to be assumed in ETAGs, ETAs and hENs (years)		
Category	Years	Categroy		
		Repairable or easily replaceable	Repairable or replaceable with some more efforts	Lifelong[2]
Short	10	10^1	10	10
Medium	25	10^1	25	25
Normal	50	10^1	25	50
Long	100	10^1	25	100
(1) In exeptional and justified cases, e.g. for certain repair products, a working life of 3 to 6 years may be envisaged (when agreed by EOTA TB or CEN, repectively) (2) When not repairable or replaceable "easily" or "with some more efforts"				

Table 1: Assumed working lives for works and construction products [3]

Standard ISO 15686 Part I (2000) [5] provides suggested minimum design lives for building components, and is presented in Table 2.

Design life of building	Components			Building service
	Inacessible or structural	Replacement is expensive or difficult*	Major replaceable	
Unlimited	Unlimited	100	40	25
150	100	100	40	25
100	100	100	40	25
60	60	60	40	25
25	25	25	25	25
15	15	15	15	15
10	10	10	10	10
Note 1: Easy to replace components may have design lives of 3 or 6 years Note 2: An unlimited design life should very rarely be used, as it significantly reducesdesign options * including below ground water				

Table 2: Suggested minimum design lives for components as in ISO 15868 (2000)

Given the above considerations, some questions arise on component or subsystem design life requirements, and especially on "non easily repairable or replaceable", or "inaccessible or structural " or "having expensive or difficult replacement" ones. The commonly used geomembranes and drainage materials, which constitute the waterproofing or watersealing and drainage systems of tunnels, will be examples of components and subsystems of such characteristics. So far with the existing knowledge, disposed statistical data, and in use application standards, we can't ensure that service life of a particular waterproofing system will be, for example, 100 years long, given the anticipated environmental conditions and the working stress regime. A rational arrangement should assume a service life for the waterproofing system less than the tunnel design life, incorporating into the design repair methods or provisions for replacement with an alternative.

Reservation and non-overstepping, at any time, of the actions and corresponding loads, which has been considered for the design of the tunnel, shall be of special concern for the design life of the tunnel. To ensure this the design has to make clear what additional loading conditions and construction methods (pilling, ground anchoring, water table lowering, excavations) are permitted and the owner has to take all provisions to fulfil legal regulations and/or expropriations.

3.2   Tunnel performance requirements

Performance requirements are the necessary attributes defined for a tunnel
before and during design. The owner's purpose is the ultimate tunnel requirement from which all other requirements originate. Requirements are
statements that identify the essential needs of the whole tunnel (tunnel does
the right thinks and doesn't do wrong thinks), in order to have value and
utility with safety, serviceability and aesthetic appearance. Performance requirements should state what the tunnel has to do, but they should not specify
how the tunnel has to do it.

Developing performance requirements is not a simple or stationary procedure,
although often-involved parties have not paid much attention to such a rational activity. Requirements are never complete neither perfect at start of the
project.

The first step in developing performance requirements is to identify anyone
who has a right to impose requirements on the tunnel. This includes owners,
funding agencies, regulatory organizations, environmental authorities, end
users, operators, emergency authorities, designers, polities, social organizations etc.

Important issue is the identification of the tunnel and what the real on hand
problem will be. This is one of the most important and often omitted items in
tunnel design, as a sophisticated solution to the wrong problem is worthless
and in some cases causes severe safety problems. For the Evinos Mornos
water supply tunnel, given the geological conditions of the area and having
constraints on the end points with specific coordinates, the statement "to
connect the two given end points" leads to a 26.5 km long tunnel. However
by adding other essential requirements the completed form of the statement
became "connect the two given end points to supply water to Athens with
considerably abbreviated construction time" leading to a 29 km tunnel that
was completed 11 months before schedule.

The characteristics of the requirements should be:

- Simple, not compound and unambiguous.

- Attainable (technically feasible, implementing into budget, time schedule and other constrains).

- Unique (not repeating).

- Accessible (if confidentiality is needed then clear classification procedures must be provided).

- Traceable back to the origin (with the relevant rationale).

- Quantitative as much as possible.

- Verifiable and testable.

The above stated requirements usually are issues of dialogue. Here we have to remind that according to ISO 8402:1994 quality is the "totality of characteristics of an entity that bear on its ability to satisfy stated and implied needs". The overlooking of such implied needs should be avoided when defining the requirements.

Quantitative expression of requirements as much as possible shall be preferable in any case. As we have already noted, there are many parts which have the right to impose requirements and usually some of them are not familiar in such way. In such cases a pertinent part has to be appointed, to translate descriptive or qualitative requirements in quantitative terms, which also have to be compatible with the design method. For example the original requirement for a rail high-speed tunnel asks for passenger comfort in the tunnel. The quantitative form asks for stringent transient pressure of no more than 3.5 kPa in 4 seconds. Whereas the designer (who has the responsibility to prove the fulfillment of the requirements) has the right to impose and to revise requirements, this duty for quantitative expression of requirements is better to be accomplished by other part (owner, independent expert etc.).

Although some of the important basic and general requirements are included in codes and regulations, it is not redundancy to resume and repeat them into the relevant document (common sense requirements should not be included). In the following we adduce cases of requirements:

- Tunnels and other underground structures shall not necessitate any structural maintenance during the whole design life. The term structural maintenance means replacement or extended repair. A more rational approach shall define the design life of the subsystems or components of the tunnel on the basis of the above paragraph 3.1. When establishing such a requirement, using an accepted principle (e.g. ALARP principle), any possible action will have to be defined and prospective exceptions will have to be exactly specified. Characteristic paradigm of exception of this requirement should be the explosion in road and rail tunnels. Many requirements for explosion accept local damage of the final lining provided progressive failure shall be prevented, and a posteriori structural maintenance is accepted.

- The owner has to specify the accepted number and duration of tunnel shutdowns for safe execution of preventative or maintenance and/or repair works.

- The tunnel shall not violate any laws or regulations.

- The owner has to define the required operational mode of the tunnel. (Fully automatic facility or with his own dedicated structure, resources and staff).

- For road and rail tunnels possible restrictions on tunnel use shall be defined. Usually for road tunnels, pedestrians, bicycles and in general pedal cycles, motorcycles with engines less than 50cc, animals and animal trailed wagons shall not permitted. Special attention and concern shall be paid to the restrictions on the carriage of dangerous goods or bulky truckloads. The owner has to decide on the admittance of such vehicles to the tunnel and the implied restrictions.

- Watertightness. For hydraulic tunnels requirements for watertightness (mostly to avoid water loss and hydraulic fracture of the rock mass) usually are formulated for acceptable water loss in liters per kilometer and per second for the operating pressure or a predetermined pressure checking level. The main disadvantage of such a requirement is the difficulty to confirm or at least to have some indications on its effectuation, in adaptable time (during the construction and before trial or full operation) and with a reasonable acceptable cost. In other words the requirement can hardly be assumed as verifiable and testable. Such quantitative requirements must be converted in other quantitative form, which can be easily tested and verified (e.g., the permeability of the system lining-joints-interfaces-rock mass). Road and rail tunnels usually require no dripping water above the road or rail surface. The rationale has to do with safety and serviceability (e.g. freezing water on road surface, falling icicles from tunnel and portal roof, flows in arcas with electrical cables and electrical equipment), durability (e.g. detrimental effects accelerating the rate of the deterioration of the structure), public acceptance of the tunnel avoiding phobia, tremor, willies and avoiding disruptive maintenance requirements in case of large water inflows.

- Acceptable impact level of the tunneling activity on adjacent and/or overhead surface structures shall have to be attained. In the past, two limit states governed this requirement:

  a) Ultimate limit state, when the structural integrity of surface or adjacent structure is jeopardized.

b) Serviceability limit state, when the designed functional use of surface or adjacent structure is disrupted.

Nowadays, the problematic public acceptance and tolerance of infrastructure projects, joint with environmental legislation constraints and time-consuming litigation procedures, may lead in adopting another limit state:

c) Cracking limit state, when cracks appear (under the acceptable width and depth for structural integrity and durability) affecting the aesthetic appearance of the structure.

The above limit states shall be translated in admissible surface deformations, which the tunneling activity must comply, regarding the type of structures, their functional use and code provisions and recommendations.

- The tunnel shall be designed to attain an earthquake resistance according to dual (two level) design principle.

  a) Level 1. Significant damages may occur. Some of the tunnel portals may be blocked, but in road and rail tunnels there will be available escape routes due to adaptable multi-escape passaways planning. The probability of collapse of a part of the final lining shall be adequately small and shall be combined with maintaining the integrity and adequate residual strength after the end of seismic sequence. The damaged structures shall be limited and repairable in defined time (usually some months). In Greece [10] suggestively the design earthquake for level 1 has probability of exceedance 5% in 100 years, corresponding to a return period ∼2000 years $\{\sim 100/\ln(1-0.05)\}$.

  b) Level 2. Minor damages may occur for the tunnels, portals and other underground constructions. Damages shall be repaired in a few days, and for that occasion tunnel closure may be acceptable. In Greece the design earthquake for level 2 has probability of exceedance 10% in 50 years, corresponding in return period ∼500 years $\{\sim 50/\ln(1-0.1)\}$.

In some cases the owner can require to design for maximum earthquake of level 1 with behavior requirements of level 2.

- The owner of the tunnel must specify the targets of optimization. In general, the life cycle cost of the whole tunnel structure shall have to be optimized, but this is not always the case.

4   Risks in tunneling

According to the above given definition, safety is absence of unacceptable (levels of) risk, therefore in studying safety the critical character of the terms "risk" and "unacceptable" is evident.

4.1   Risk

Risk has many definitions depending on the discipline and the consideration of the involved individual. Risk R in general can be defined as an action that jeopardizes something of value. The risk (of an activity with only one hazard with potential consequences C) may be calculated as the product of the probability P that this hazard will occur during a stated period of time, and the consequences given the hazard occurs. In symbolic quantitative terms we can write: $R = P \cdot C$.
Usually the above stated period of time coincides with the design life of the tunnel, and the risk can arise at any time within that period.

To describe the overall risk of a multi-hazard activity, the risk associated with each hazard should be summed up for all hazards. That single number should be meaningful only with the prerequisite of a weighting procedure, which necessarily involve value statements. We have to emphasize that a single number is not sufficient to describe precisely and in whole the real risk, but it can provide a sufficient basis for risk comparison, for a given situation. The above definition often conceals significant information interesting for the decision makers. To overcome this constrain the risk has to be seen as the answer to the next fundamental questions:

- What can go wrong?

- What is the probability or frequency?

- If it occurs, what are the consequences?

According to this approach, risk has three elements:

- Scenario (hazard identification)

- Likelihood

- Severity

Of great importance in risk definition and risk management are the boundaries of the risk domain. A specific road or rail tunnel has to be considered as a separate infrastructure, or the transport system has to be seen as a whole?

4.2 Risk acceptance criteria

Risk acceptance criteria are the qualitative or quantitative expressions defining the maximum risk that is acceptable or tolerable for a given system. Fischhoff et al. [11] claim that no risk is acceptable unconditionally. Strictly speaking, no risk is acceptable unless, as a principle, risk can be considered tolerable when there are sufficient compensational benefits. Accordingly, safety does not postulate all risk to be extinguished, but daresay claim for an appropriate balance between risk, cost and benefit. From that point of view, acceptable risk problem is a decision problem, choosing between alternatives.

Risk societies accept or tolerate a level of risk. The evident or alluded background of risk acceptance criteria can be summarized below:

- It is advisable to have risks with large probabilities and small consequences rather than risks with small probabilities and large consequences (tragedies).

- The higher acceptable risk the higher the resultant benefit.

- Risks that can be demoted or eliminated by reasonable means, should always be excluded.

- Individuals or groups (or societies) should not experience risks that are not in proportion of the received benefits.

4.2.1 Risk acceptance criteria to be used in quantitative methods

When dealing with risk acceptance, in assessing risk with quantitative methods, we must distinguish the subject of the risk; so we have to consider individual risk and risk in society or groups.

4.2.1.1 Acceptable individual risk

At first consider an individual who undertakes an activity balancing the risks against resultant benefits. Acceptable individual risk is defined as the frequency at which a specific individual accepts to sustain a given level of harm or damage due to the realization of specified hazards, with account taken of time aspects. The "given level of harm or damage" is usually restrained to the loss of life, and is expressed in a proper way that considers the metric of death.

The matter of what risks can be accepted is a very attitudinal issue, that strongly depends on the preferences of the specific individual. The above definition of acceptable individual risk is only one from in use definitions on this item. It is self-evident to use the same definition for individual risk when comparing relevant risks. Also it is crucial to ensure the compatibility of the used units, as different units for the same risk can result in an absolutely different understanding of risks. Individual risks deduced from statistics of causes of death and the number of participants involved into activity. Factors affecting the magnitude of acceptable individual risk can be summarized as below:

- Self-control of the individual in the activity.

- Voluntarism (as a crude rule public tolerates 1000 times greater risks from voluntary than from involuntary activities).

- Perceived benefit (direct or indirect).

The basis for acceptable individual risk tends to be the de minimis value 10^{-6} deaths per year, which indicates a low acceptable risk level, where everybody can live with. This value is set for involuntary imposed risks (e.g. for individuals living near the tunnel).

Jonkman et al. [14, 15] have proposed a criterion for acceptable individual risk for tunnels (IR), which takes into account the voluntarism of the activity and the benefit perceived:

$$IR < \beta \cdot 10^{-4} (yr^{-1})$$

Where: β is the policy factor, which varies according to the degree of the voluntarism of the activity, and the benefit perceived. Jonkman et al. have proposed values (see Table 3) for the policy factor.

The above criterion is based on the degree of the voluntarism and the benefit perceived. Another important factor, the degree of personal influence on the success of the activity, is not included there. This factor should be used for the accepted individual risks during the design and construction of the tunnel. According to this approach, and based on Bohnenblust (1998) [17] four risk categories for individual risk can be formulated, as in Table 4.

In UK HSE's document R2P2 [19] suggests tolerability limits for individual risk (individual death). Risks greater than 10^{-3} per year for workers and

Party	β
Employees (rail)	1
Passenger or user	0.1
Person living near the tunnel	0.01

Table 3: Proposed β values for different parties involved in tunnel safety. (Jonkman et al. 2003) [14]

Risk categories	Individual risk. *Highest value of probability of death per year*
Category 1: 100% voluntary	$10^{-2} - 10^{-3}$
Category 2: large degree of personal reponsibility	$10^{-3} - 2 \cdot 10^{-4}$
Category 3: small degree of personal reponsibility	$2 \cdot 10^{-4} - 3 \cdot 10^{-5}$
Category 4: 100% involuntary	$3 \cdot 10^{-5} - 4 \cdot 10^{-6}$

Table 4: Individual risk categrories

10^{-4} per year for the public shall be regarded as unacceptable, whereas risks smaller than 10^{-6} per year, are broadly acceptable. In UK and from Railway Group Safety Plan 2003-2004 [18] and HSE's document R2P2 [19] criteria are given for the individual passenger risk, individual employee risk and individual public risk, as shown in Table 5.

According to the aim of the analysis, formulation of individual risk criteria can be done in several types. For example in road tunnels we can use the following individual risk criteria:

- Risk is unacceptable if it is higher than 10^{-x} per year.

- Risk expressed in fatalities per million trips is suggested as a measure for risk reduction in case of a tunnel on an existing old road.

- Fatality frequency on the road should not be more than y fatal accidents per 100 million vehicle km. Considering the average number of passengers per vehicle the above form can be converted to fatal accidents per 100 million person km.

Group	HSE upper limit of tolerability (probability of fatality per year)	Railway Group Safety Plan 2009 targets (probability of fatality per year)	HSE broadly acceptable (probability of fatality per year)
Individual passenger risk	10^{-4} (1 in 10,000 per year)	1 in 133 million passenger journeys, i.e. $3.38 \cdot 10^{-6}$	10^{-6} (1 in 1,000,000 per year)
Individual employee risk	10^{-3} (1 in 1,000 per year)	zero fatalities per year	10^{-6} (1 in 1,000,000 per year)
Individual public risk (person living near railway)	10^{-4} (1 in 10,000 per year)	10^{-6} (1 in 1,000,000 per year)	10^{-6} (1 in 1,000,000 per year)

Table 5: Individual risk criteria for UK Railway

4.2.1.2 Acceptable societal risks

Societal risk is defined as the risk to a group of people due to all hazards arising from a hazardous operation. The Institute of Chemical Engineers (1985) [21] has adopted the following definition for societal risk; "the relation between frequency and the number of people suffering from a specified level of harm in a given population from the realization of specified hazards". The objective of societal risk criteria is to abridge the risk for groups, such as local communities, or the whole society. Societal risk terms provide much more detailed information about the risk nature than individual risk; especially how many fatalities can be expected every year and how many fatalities should be expected in occurrence of a hazard. Individual risk gives the probability to die in a certain location (local property), whilst societal risk gives the number of fatalities for a major area, independently of the precise origin of the hazard.

If the "specified level of harm" in the above definition of societal risk, restrained to the loss of life, the most common, suitable and convenient way to express and to present societal risk is the graphical type of FN curve (or Farmer diagram). Note that the FN curve is also applicable for presentation of individual risk. This curve represents (usually in a double logarithmic scale) the frequency $F(N)$ of accidents as a function of the number N or more fatalities. We stress the cumulative type of consequences denoted by the expression of N or more fatalities. The FN curves always are falling from left to right or flat. To proof this we consider the relation:

$$F(N) - F(N+1) = f(N)$$

Where $f(N)$ is the frequency with exactly N fatalities Given that always $f(N) \geq 0$ we extrapolate that always

$$F(N) \geq F(N+1)$$

The point where the FN curve crosses the vertical axis, which is located at $N = 1$, is the frequency of accidents with 1 or more fatalities $F(1)$ or the overall frequency of fatal accidents.

The surface under the FN curve represents the expected value of the number of fatalities.

From the FN curve one can transparently conclude the risk of having a certain number of fatalities in a single hazard which an individual risk can not do. Units for the frequency (or probability) must be selected carefully, to accommodate the scope of the risk management. E.g. when the analysis for road tunnels, focused on the risk as an object for the society we should use the annual frequency in the tunnel, whilst the risk used for comparative studies should use frequency per million person km.

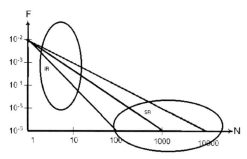

Figure 1: Frequency F of N or more fatalities per year.

Figure 1 is an illustrative representation of FN curve. Tunnel accidents of societal concern usually are events with small probabilities and large consequences. In Figure 1 the SR area corresponds to societal risks, and the IR area corresponds to individual risks. Road tunnel collisions have to be treated as individual risks, while fires, explosions or release of toxic substances have to be treated as societal risks.

The straight lines in Figure 1 represent the "acceptance criteria". The area above each line represents the region of unacceptable risk for the corresponding criterion. The decreasing public tolerability to the large fatalities risks, called risk aversion, is represented by the lower lines.

Prescribing a societal acceptance criterion, e.g. a line on the FN diagram, it is necessary to assign a fixed point, the gradient of the line, and in some cases cut-off values of consequences and/or probabilities.

Fixed points have been derived by an assortment of methods, which include analytical (which pay no regard to cost-benefit effect), expert judgment, boot-strapping to existing societal risk acceptance criteria. To date there is no universally accepted method to identify fixed points. It is preferable the fixed points to apply to the upper limits of tolerability. Synoptically used values for fixed points in UK and Hong Kong are $(F : 10^{-4}, N : 10)$ and in the Netherlands $(F : 10^{-5}, N : 10)$ [24].

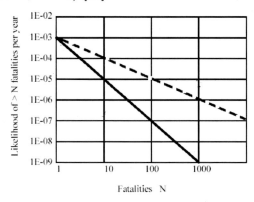

Figure 2: Societal risks; dotted line UK and Hong Kong, straight line Netherlands.

The most societal risk criteria can be written in the form:

$$F \cdot N^a = C$$

Where F : probability of N or more fatalities
       N : number of fatalities
       a : "aversion" index (1 or 2)
       C : constant positioning the FN line

The values of a and C can vary over the N axis. The gradient of the line representing societal risk criteria, plotted in a double logarithmic scale, is $-a$. When the value of a is 1, as in UK and Hong Kong, then we speak of neutral risk. Correspondingly to a value of a 2, as in the Netherlands, then we speak of risk aversion.

In the UK HSE's document R2P2 [19] suggests for upper limit of tolerability for multiple fatalities due to an existing installation with more than 50 fatalities, the value of $2 \cdot 10^{-4}$ (1 in 5000) or 10^{-5} to 10^{-6} for new installations.

The above line in FN diagram represents the border of unacceptable region. A particular risk falling into the region above the line is regarded as unacceptable whatever the level of the resultant benefits is.

Having defined the unacceptable region on the FN diagram, we can also define the border line of a broadly acceptable region. A particular risk falling into this region, below the defined border, is regarded as insignificant and adequately controlled; no further action is required to reduce it unless reasonably practicable measures are available. The area between the unacceptable and broadly acceptable region is the tolerable region. ICOLD (2002) [22] has defined tolerable risk as "a risk within a range that society can live with so as to secure certain net benefits. It is a range of risk that we do not regard as negligible or as something we might ignore, but rather as something we need to keep under review and reduce it still further as we can". The most common vehicle to qualify tolerable risk is the ALARP (as low as reasonably practicable) principle. ALARP is implied straightly in the above definition of tolerable risk.

4.2.1.3 Setting up acceptable risks for tunnels

Under conditions, quantitative methods of analysis can be considered as being the most objective ones. To redeem the attitude of objectivity more time and resources are required. Having quantified probabilities and consequences it is possible to establish risk acceptance criteria for the hazard considered.

There are several alternatives for developing quantitative risk acceptance criteria, like:

- Comparing with statistics from existing tunnels, averaging a risk level (e.g. comparing with risks accepted previously)

- Comparing with general risks in society and individuals

- Comparing with risks from other hazardous activities, for which there are well-established acceptance criteria

- Using mathematical approaches

- Adopting specified regulations and/or codes

For the present, adequacy of relevant statistics and risk analysis of existing tunnels is unambiguously questionable. To determine probabilities of accidents (hazards) large and detailed historical data is required. Register and insertion in a database has to be done in a standardized mode, taking into account that the probability can be separated in the probability of occurrence and the probability of detection.

For traffic accidents normally good statistics are available, and if we have good statistical data then we should use them. History data show the disposition of society to maintain the current risk level. Probably the most reasonable way to formulate societal criterion in the FN diagram should be a combination of good statistics, expert judgment, and target preferences. The first step defines a point in the FN acceptance curve. The second and maybe more questionable step is the slope of the curve. To date the existing societal risk criteria have been derived for the evaluation of societal risks associated with hazardous industrial activities (nuclear plants, chemical plants), airports. There are little available data for tunnel-associated hazards and risks resulting in problematic and unreliable estimation of parameters. From that point of view one has to take into account the differences between the considered installations safety and risk and the tunnels safety and risk. The main difference is that the tunnel users joint the activity with a degree of voluntarism and at the same time benefit from using the tunnel. Therefore the acceptable societal risk for road tunnel users can be higher by an order of magnitude than for the above-mentioned installations. Regarding the slope of the FN curve, as was mentioned above the corresponding values are typically -1 (risk neutral) and -2 (risk aversion). Synoptically we make some references about risk aversion and the implementation in deriving societal risk criteria. Risk aversion is an "over reaction" based on past experience and is, thus, a matter of information. Risk aversion can be reduced or even eliminated considering all important hazards into the equation for the total expected benefit. Based on that approach, and judging the level of the "in-formed" character, the slope of the FN curve can vary in the interval -1 to -2, with preference to subinterval -1 to -1.5. Based on past experience (Salag tunnel accident) a cut off value of 1000 fatalities should be inserted.

4.3   Risk management

Risk management is referred to systematic techniques, which are used throughout the evolution of tunnel project. Risk management includes risk analysis (identification of hazards and description of risks), risk evaluation, risk assessment, risk elimination and risk mitigation measures.

4.4 Risk Analysis

Risk analysis has been defined as a structured process that identifies both the probability and extent of adverse consequences arising from a given activity, including hazard identification and description of risks, in qualitative or quantitative way.

4.4.1 Hazard identification

Hazard identification is the first critical step in risk management procedure, related to the first question which has to be answered: "what can go wrong?" Hazards have to be recorded at any subsystem and at any point in tunnel life cycle, starting with early design stages. Although there can be no assurance of discovering all hazards, identification of hazards has be done in an organized and structured way.

Hazard identification and hazard discovering must follow some rules and proven methods:

- Using the principle "four eyes are better than one", finding hazards should be a team's responsibility. Brainstorming also is useful when applied effectively.

- Review similar tunneling projects, recorded in literature.

- Use instinctive engineering percept.

- Review codes, regulations, and standards.

- Review performance requirements.

- Review the on use and future management scheme. Look for possible problems, specially for communication and managerial problems.

- Review the proposed method and technology, especially the familiarization of all involved parties.

- Consult checklists.

- Check for energy sources and procedures for harmful release.

- Consider all stages of tunnel implementation, from the conceptual phase to the detailed design, auditing, acceptance of the tender, construction, insurance, reinsurance, commissioning, operation and decommissioning.

- Identify as many hazards as possible.

- Discuss with qualified and experienced staff from design and/or construction responsible. Use experts' opinion.

- Organize hazards in groups. Eskesen et al (2004) [13] has proposed two major groups: General hazards that may be considered generally for each contract, and specific hazards that should be considered for each part of the project.

  General hazards:

  1. Contractual disputes

  2. Insolvency and institutional problems

  3. Authorities interference

  4. Third party interference

  5. Labor disputes

  Specific hazards:

  6. Accidental occurrences

  7. Unforeseen adverse conditions

  8. Inadequate designs, specifications and programs

  9. Failure of major equipment

  10. Substandard, slow or out of tolerance works

- Don't confuse a hazard with its consequences. The formulation should follow the sequence: Source – Mechanism – Consequence (suggestively: don't use "Methane Explosion Damage". We recommend "Methane concentration from the surrounding rock mass in presence of ignition sources – Detonation – Injuries/Fatalities/Equipment and/or structure damages")

- Look into the possibility for application of formal approaches that can be used to identify hazards, e.g. Preliminary Hazard Analysis (PHA), System Hazard Analysis (SHA), Interface Hazard Analysis (IHA), Operating and Support Hazard Analysis (O&SHA), Hazard Indentification (HAZID), Hazard and Operability Studies (HAZOP), Failure Mode and Effect Analysis (FMEA). Although all of these structured, organized and auditable approaches have been developed for other sectors of engineering, they can be used in tunneling without many modifications and adjustments.

4.4.2 Description - identification of risks

After identifying the hazards, that is to say answering the first question on what can go wrong, the next step is to identify the related risks. Identification of risks comprises the last two elements of risks: their probability, and their consequent severity. Analyzing risks, both the magnitude of probabilities and of the consequences are of importance. Treatment of these elements can be done using qualitative, quantitative or a hybrid method, but always has to be done in accordance with the risk acceptance criteria defined for the tunnel.

A. Probability

Probability or frequency of occurrence can be scored qualitatively or quantitatively. Suggestively, for describing probability, the most common in use means are:

- Look into experience of all the concerned parties and similar tunnels.

- Simulations (physical models, numerical and/or analytical simulations).

- Engineering (expert) judgment using subjective scoring scales.

- Quantitative probability expressions.

The following restricts preference for quantitative approaches:

- Quantitative statistics are only in a few instances available, as extensive historical data is required. The number of events (hazards) may often be low, far too few to make reliable predictions.

- Quantitative approaches are based on occurrences in the past. For some types of hazards, changes in causal factors have been confirmed, which are not reflected in historical data. Aging of tunnels, reduced maintenance, improving vehicles, different codes and regulations may be considered as a causal factor.

- Register and insertion in a database has to be done in a standardized mode, taking into account that the probability can be separated in the probability of occurrence and the probability of detection.

Because of the above restrictions, probability description is often done in a qualitative way. Engineering (expert) judgment is the most common approach to describe probability. This expert judgment should be a team work where members contribute their own experience in a structured and evidential way. Expert judgment should not been considered as magisterial secret.

As every tunnel has its own specific characteristics, qualitative description of probability (or frequency) should be tunnel oriented and tunnel targeted. Nevertheless some rules of thumb should be taken into account:

- No expert judgment can disaggregate more than six probability steps.

- The more discrete the probability steps are probably the more confusion we have.

- Separate adjacent probability steps by a factor of 10.

- When considering probability, the time period for which the hazard has been identified (e.g. construction period, operation period) is of special interest.

Eskesen et al. (2004) [13] generally recommended a separation into five classes as a practical way of classifying frequency. Table 6 shows an example of recommended classification in the construction period.

Frequency of occurence (in the construction period)			
Frequency class	Interval	Central value	Descriptive frequency class
5	>0.3	1	Very likely
4	0.03 to 0.3	0.1	Likely
3	0.003 to 0.03	0.01	Occasional
2	0.0003 to 0.003	0.001	Unlikely
1	< 0.0003	0.0001	Very unlikely

Table 6: Frequency of occurrence in the construction period (after Eskesen et al. 2004) [13]

As it is mentioned above, qualitative description of probability (or frequency) should be tunnel oriented and tunnel targeted, so the recommendations are not inviolable rules. Especially when little data and information are available, as in the early stages of the design, it is better to elaborate a classification scheme with three or four frequency classes. Table 7 illustrates an example of three separation classes.

Frequency of occurence (in the construction period)	
Descriptive frequency class	Description
Likely	More than one occuring
Occasional	At least on occuring
Remote	Unlikely to occure

Table 7: Three separation classes for frequency

B. Consequences

Like probability classification, consequences have to be treated individually for each tunnel. The consequences to be considered in risk analysis should be the same as defined in acceptance criteria.

Types of consequences can be categorized as follows:

- Adverse consequences to human life or health. Based on the explicated concepts of individual and societal risk, further differentiation is recommended:

    - Fatalities or/and injuries to tunnel personnel and emergency crew.
    - Fatalities or/and injuries to third parties (people living near tunnel, passengers).

- Damage to tunnel structure and/or subsystems.

- Damage to other structures (aboveground and/or underground).

- Harm to environment.

- Delay to time schedule of the tunnel.

- Inobservance to fulfill predefined (functional and other) requirements that results in economic loss to owner (or to appointed party).

- Loss of reputation.

Just like probability, determining all the above consequences in a quantitative way (e.g. specifying economic losses, extra costs, loss of requirements, environmental impacts, injuries or loss of lives) is a difficult issue. Usually consequences are classified into some classes or intervals. The number of consequence classes for the most cases, varies from four to six. For each of the above consequence types (categories), consequence classes shall have to be defined. Elaboration of consequence classes shall have to result in consistent and manageable and where is possible in unified classes. Describing consequence classes in a quantitative way, separation of adjacent classes in general is recommended to be done by a factor of 10. For some types of consequences (e.g. delay) applying such a rule may result in meaningless and unmanageable consequence class. Eskesen et al. (2004) [13] proposed a five-class consequences classification system, which use unified class descriptors for the most of the above consequence types, see Table 8.

Description	Consequence types					
	Injury to workers & emergency	Injury to Third parties	Damage to Third party property (loss per hazard in Mil. €)	Harm to the environment	Delay (in month per hazard)	Economic loss to owner (in Mil. €)
Disastrous	$F > 10$	$F > 1$, $SI > 10$	> 3	Permanent severe damage	> 24	> 30
Severe	$1 < F < 10$, $SI > 10$	$1F$, $1 < SI \leq 10$	$0.3 - 3$	Permanent minor damage	$6 - 24$	$3 - 30$
Serious	$1F$, $1 < SI \leq 10$	$1SI$, $1 < MI \leq 10$	$0.03 - 0.3$	Long term effects	$2 - 6$	$0.3 - 3$
Considerable	$1SI$, $1 < MI \leq 10$	$1MI$	$0.003 - 0.03$	Temporary severe damage	$0.5 - 2$	$0.03 - 0.3$
Insignificant	$1MI$	–	< 0.003	Temporary minor damage	< 0.5	< 0.03

Table 8: Consequence classes for different consequence types (after Eskesen et al 2004) [13] (F: fatalities, SI: serious injury, MI: minor injury)

C. Description of Risk

Each hazard risk can be identified by taking into consideration its elements: Probability (or Frequency) and Consequences. If both probability P and consequences C are in quantitative form, risk can be quantified as the product P times C. Unfortunately such an approach is too simple to be applicable. In fact, simplicity is based on the ignorance of uncertainties when estimating probabilities and consequences. To overcome this disadvantage, description of uncertainty should be incorporated into the quantitative estimation of probabilities and consequences. Describing methods and techniques for modeling uncertainty in engineering problems lies beyond the scope of the present paper. When selecting methods and techniques for uncertainty modeling of a specific hazard, their applicability has to be checked on the basis of the nature of the hazard (low probability – high consequences hazard or high probability – low consequences hazard).

In case of qualitative identification of probability and/or consequences, a semi-qualitative or qualitative description of risk has to be carried out. The most common method is the risk matrix, which combines probability versus consequences. As long as risk acceptance criteria must be defined for a certain tunnel and risk matrix shall be compatible with risk acceptance criteria, risk matrices have to be defined for each certain tunnel. Risk matrix has to approximate acceptance criteria with matrix cells.

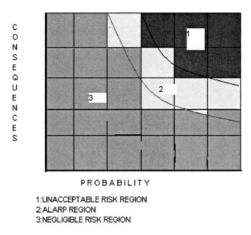

PROBABILITY

1:UNACCEPTABLE RISK REGION
2:ALARP REGION
3:NEGLIGIBLE RISK REGION

Figure 3: Fitting matrix cells to acceptance risk criteria boundaries.

Apart from compatibility with risk acceptance criteria, risk classification or risk matrix zoning have to be designed properly corresponding to the intended use. Although there are many risk matrices that have been developed and used, each tunnel challenges the development of its specific risk matrix. In order to ensure effective use of a risk matrix, some rules have to be satisfied:

- Spell out how the matrix is intended to be used. The risk matrix has certain limitations and can't be considered as a nostrum.

- A risk matrix must have at least three regions corresponding to unacceptable risk, tolerable (e.g. ALARP) risk, and accepted or negligible risk. Risk acceptance criteria must be transferred and implemented carefully on the risk matrix.

- Developing of too many zones and risk categories is only adding confusion without increasing the risk matrix effectiveness. In most risk matrices four risk categories have been established, and regardless of used terminology these can be defined as follows:

    - Category I (Unacceptable): Risk is absolutely unacceptable and shall be reduced to a lower risk category regardless of the costs.

  – Category II (Unwanted, Undesirable): Risk shall be reduced by mitigation measures as long as the costs of measures are not proportionate with the risk reduction obtained and the outcome is tolerable.

  – Category III (Acceptable): Risk is tolerable and shall be managed throughout the project without further consideration of mitigation measures.

  – Category IV (Negligible): Risk is acceptable without further consideration.

- A risk matrix must be organized in such a way so that it is having the capability to assess the influence of mitigation measures.

- When using only one matrix with multiple consequence classes, compatibility of all types of consequences must be ensured.

- Make clear that probability is the probability with which the hazard occurs and not the probability with which the consequences emerge. Take into account all interdependencies of the probability with which the hazard occurs and the probability with which the consequences emerge.

4.5   Risk elimination and risk mitigation measures

In evaluating risks there are four options:

- Accept risk without any further consideration.

- Transfer risk to someone else who is minded to accept and is legally admissible.

- 'Eliminate risk' means to prevent risk occurrence (e.g. eliminate the first element of risk that is risk scenario). Typical examples: Frontal collisions can be avoided by constructing twin tunnels that can separate traffic directions. Some geotechnical risks can be avoided by changing the alignment or constructing a bridge instead of a tunnel.

- Mitigate risk by reducing the probability of occurrence or consequences.

Regarding safety the latest two options may be of importance.

Usually direct comparison of the products of a risk analysis with risk acceptance criteria provides a sound basis for decision making on risk elimination and risk mitigation measures. The residual risk remaining after application of mitigation strategy and activation of mitigation measures can be assessed in the same way for determining acceptability.

Secondary risks, which arise from actions taken to mitigate other risks, or from extensions to the original scope of the project, have to be assessed in the same cyclic mode.

In a risk elimination and risk mitigation process, prioritizing risks should be considered as a very important issue, especially for risks in ALARP region. Such a structured process can identify the risks that are the most critical to the tunnel and, therefore the most in need of actions to be taken.

Another key point in selection of risk mitigation measures is the effectiveness of these measures, taking into account the refund of the related investments. Jonkman et al. [14] proposed a framework to analyze cost effectiveness of risk mitigation measures. The main problem in working out such an optimization analysis is the lack of knowledge on the quantified effects of risk mitigation measures and the interactions of various risk mitigation measures. Hence and taking into account cumulative effects of various risk mitigation measures, rational absolute optimization of cost effectiveness of risk mitigation measures is not feasible. However relative comparisons of the quantitative effects of individual risk mitigation measures can be done, leading up to a full probabilistic optimization analysis of risk mitigation measures.

Risk mitigation strategy and selection of mitigation measures have to be considered over a certain period of time, similar to the one used in risk analysis. Risk mitigation strategy has to ensure that it still mitigates the risk effectively over that period, so monitoring and review acceptability of residual risks have to be implemented into mitigation strategy. Risk mitigation may come up from reducing one or both of the latest risk elements; probability or/and consequences. In a broad sense risk mitigation measures should fall into two categories: measures reducing probability and measures reducing consequences. Undoubtedly risk mitigation measures serve in a multipurpose way and can therefore not be considered as acting exclusively in an "or" mode.

4.5.1 Risk mitigation measures in the operational phase of road tunnels

Despite the problematic division of risk mitigation measures we can distinguish measures reducing probability and measures reducing consequences.

4.5.1.1   Measures reducing probability

A. Tunnel geometric design

Most national guidelines e.g. [10] and directive 54/2004 [25] provide an adequate background for the tunnel geometric design. Items of geometric design that may have remarkable effect in risk mitigation are:

- Tunnel cross section

- Single or twin tube tunnel

- Horizontal and vertical alignment

- Drainage

- Escape routes

- Manageable assess for emergency personnel (and vehicles)

- Hard shoulders

- Emergency lay-bys

- Turnover layout

- Finishes of tunnel surfaces (road, walls)

- Layouts confining fire and/or smoke spreading (e.g. flame traps, pressurized parts of tunnel system)

- Shape and layout of tunnel portals.

- Adoption of the alignment so as to manage a uniform reduction of operational velocity from open road to inside the tunnel section.

B. Operational mode and traffic regulations

As long as there is a system that can enforce and ensure observance the following means can reduce the probability of an accident:

- Written procedures for execution of maintenance or repair works.

- Organizational layout according to [25].

- Ensuring intended use of tunnel and fulfillment of restrictions on tunnel use. Signs must prohibit entrance of non- anticipated traffic (e.g. pedestrians, bicycles and in general pedal cycles, motorcycles with engines less than 50cc, animals and animal trailed wagons). System for height control and adaptable layout for alternative route.

- Special attention and concern shall be paid to the restrictions on the carriage of dangerous goods or bulky truckloads and the possible requirements for realization (e.g. special parking areas, or escort vehicles).

- Traffic regulations (e.g. speed limit, prohibition to overtake, minimum distance between vehicles).

C. Electromechanical equipment

- Lighting system designed, constructed and checked according to guidelines [26, 27, 28].

- Ventilation system. From the perspective of reducing probability, ventilation system has to maintain adequate visibility into the tunnel and remove or reduce concentrations of harmful substances contained in exhaust gases from vehicles (or rarely from some natural fermentations).

- Signals and information system.

- Pumping system.

- Back-up and emergency power supply system.

D. Durability design of final lining
The design has to ensure durability for the entire design life of the tunnel. Quality assurance procedures have to ensure workmanlike implementation of the foreseen measures.

E. Maintenance and inspections
The role of maintenance and inspections in reducing probability of accidents is crucial. Maintenance should be done according to written and approved procedures by authorized personnel. Also early detection of deviations from design assumptions may be of great importance to prevent unconsidered states.

4.5.1.2   Measures reducing consequences

A. Choice of rational design scenarios

Choice of design fire plays an important role in designing ventilation system, fire detection system, escape routes, structural design of final lining and fire resistance of equipment and systems.

Design fire for ventilation system, escape routes and structural design of final lining shall be represented in most circumstances by a burning lorry with a maximum heat development 30-100 MW. Although a fire power of 30 MW has been recommended [29] as basis for dimensioning the ventilation system in tunnels, a tendency for increasing at least up to 100 MW has been evident. Larger fires, like tanker carrying 50 m^3 gasoline (300 MW), are not recommended as design fires, mainly because current technology is not capable of facing up with such a huge event. In such circumstances reduction of probability appears as the most feasible mean.

Sprinkler systems shall be designed for fires of about 10 MW while fire detection system for smaller fires representing passenger car fires for 3 MW.

B. Emergency equipment

- Fire extinguishers

- Permanent water supply-hydrants

- Fire station (water or foam type fire plugs)

- Sprinklers and other automatic extinguishing systems (e.g. water barriers, curtains)

- Automatic fire detection system

- Automatic detection system for traffic incident, gases and visibility measurement, vehicle identification system, speed detection system

- Emergency lighting

- Emergency telephone

- Loudspeaker (intercom)

- Closed circuit television system

- Video surveillance

- Radio re-broadcasting equipment

- Traffic control center

- Entrance information system

- Signals and other information systems (emergency exit signal, blinking red stop signal, remote controlled barrier variable sign, driving lane indicator, service sign)

- Tunnel management center

C. Emergency action plan

A well-made emergency action plan that have to be implemented by well-trained, suitably equipped and authorized emergency services is a must for every tunnel.

4.6 Risks in risk management

Traditionally, safety should be obtained with the conformity to codes, standards or regulations, and a system to put all of them into practice (e.g. inspections). Such conformity scarcely assures against co-existing and/or inherent blemishes and related failures of individual components of the tunneling system. We have to admit that code, standard, regulation conformity is necessary but not sufficient.

Risk management seems to be the rational step to render the whole procedure sufficient. Anyway, even that procedure will be susceptible to some risk that can impair the contribution of risk management. Such risks may be as follows:

- Inadequate resources (e.g. time, money, information, and knowledge) can cause ineffective risk management.

- Improper mixture of experiences of a risk management team. Risk management teams must have an interdisciplinary character.

- Composition of risk management team in improper time. The team has to be activated as earlier as possible.

- Inappropriate risk management procedures and methods can cause confusion and pass over the real problem.

- Loosely defined interfaces and interdependencies between involved parties. (e.g. owners, designers, operators, contactors, consultants, auditors, authorities etc.)

- Risk on oversimplification and uncritical trustfulness on codification. Whereas this motivates the use of formalized risk assessment procedures, on the other hand it may restrict the view in addressing risks [30]. The HSE report on the Heathrow tunnel collapse focuses on such hazard with its observation "the pro forma approach to risk assessment kept the focus on routine worker safety. It did not encourage the strategic identification of high-level engineering and management issues essential to the success of NATM, such as the major hazard events and their prevention" [31].

- Uncertain and ruleless procedures into risk management team.

- Not well-defined or ambiguous risk assessment strategy. Risk management team members have to adopt a common convention when assessing consequences. The most rational approach is to consider worst credible consequences. (Avoid most probable or worst conceivable consequences)

Stefan Tietz [34] considered that the use of formalized risk management procedures has become fashionable and argued that they can jeopardize when being exacting and overestimating. His opinion can be briefly summarized as "Formalized risk assessment and evaluation methods should be used as an aid to professional judgment, not as a substitute for it. Professional judgment is by far the most important tool in risk management". This is partly true. In order to accomplish a systematic risk management is required a preadaptation in carrying out such activities, a bare minimum familiarization with risk management concepts and an adaptable well-equipped organizational structure. Remember also that the sooner the activation of risk management procedures will be, the better we can benefit from the effectiveness of risk management. Fashionable (or not) risk management, when should be used in a conscious manner, in any case will be preferable than no risk management.

5   Safety during construction period

5.1   Legislative framework

Relatively recent developments in the legal framework concerning construction have had a serious influence on execution of tunneling works. This legal framework is oriented basically in defining and managing technical and organizational risks that are directly related to safety. In most European countries European Community Directive 92/57/EEC of 24 June 1992 on the implementation of minimum safety and health requirements at temporary or mobile construction sites (eight individual directives referring to article 16(1) of EEC Directive 89/391) has become the basis for the legal framework that regulates tunneling works [32, 33]. The Directive 92/57 represents a novel approach for the safety and health in construction sites, based on the principle of contribution of all those involved in the construction activity. Implied aftermath of that principle was the concept of shared responsibilities of all involved parties (owners, designers, contractors, supervisors, working staff). All these are responsible, in their respective area of involvement and expertise, for their own safety and health but also for safety and health of anyone else who may be affected by their actions or decisions. This has established a new safety philosophy, setting aside traditional approaches that safety and health on construction sites was responsibility only of the contractor. The Directive also introduced new safety and health documents like prior notice, safety and health plan and safety and health file while safety and health coordinators designated for the design phase and the construction phase.

As a prerequisite for the execution of duties, safety and health coordinators must be deeply aware of the "General Principles of Prevention" [33, clause 6] These supreme principles can be summarized below:

- Avoid risks.

- Evaluate the risks which cannot be avoided.

- Combat the risks at the source.

- Adapt the work to the individual, especially as regards the design of work places, the choice of work equipment and the choice of work and production methods, with a view, in particular, to alleviate monotonous work performed in a predetermined work-rate and to reduce their effect on health.

- Adapt to technical progress.

- Replace the dangerous by non-dangerous or less dangerous.

- Develop a coherent overall prevention policy which covers technology, organization of work, working conditions, social relationships, and the influence of factors related to the working environment.

- Give collective protective measures priority over individual protective measures.

- Give appropriate instructions to the workers.

Safety and health coordinators have to interpret the above principles in the light of tunnel design and/or tunnel construction for each tunnel.

Safety and health coordinators during the design phase or the construction phase have well defined duties according to the legislation. Duties of safety and health coordinators during the design phase are summarized below:

- Coordinate the implementation of the general principles of prevention when: (1) architectural, technical and/or organizational aspects are being decided and (2) estimating the period required for completing the project phases.

- Draw up, or cause to draw up, a safety and health plan setting out the rules applicable to the construction site concerned.

- Prepare a file appropriate for the characteristics of the project containing relevant safety and health information to be taken into account during any subsequent works.

Usually, concerns of a safety and health coordinator during the design phase will be:

- How will each subsystem of tunnel be built without exposing people in a dangerous situation that can affect their life and health?

- Are there other technical solutions and methods that are less dangerous and fulfil the performance requirements?

- How will each subsystem of tunnel be maintained and/or repaired during the design life, without exposing people in a dangerous situation that can affect their life and health?

- Which materials will be used for the tunnel construction and how can they affect people's life and health?

Correspondingly the duties of safety and health coordinators during the construction phase are summarized below:

- Coordinate the implementation of the general principles of prevention when: (1) technical and/or organizational aspects are being decided and (2) estimating the period required for completing such work or work phases.

- Coordinate the implementation of the relevant provisions in order to ensure that employers as necessary for the protection of workers or self-employed persons: (1) apply the principles (specific on construction sites) in a consistent manner and (2) where required, follow the safety and health plan.

- Make, or let be made, adjustments required to the safety and health plan and the safety and health file to take account of the progress of the work and any changes that have occurred.

- Organize cooperation between employers, including successive employers on the same site, coordination of their activities with a view to protect workers and preventing accidents and occupational health hazards and reciprocal information as provided in the framework Directive [33], ensuring that self employed persons are brought into this process where necessary.

- Coordinate arrangements to check that the working procedures are being implemented correctly.

- Take the steps necessary to ensure that only authorized persons are allowed to enter the construction site.

It is important to note that the Directive's provisions apply to all construction sites, above ground and underground. However, due to the complexity and the special nature of tunneling works, the used methods have to support all parties to conceptualize interdependencies and interactions between design and construction and the resultant safety level. On account of the aforesaid sentence the temporary conditions during construction of tunnels require more in-depth consideration by design than the case of other above ground civil engineering works.

5.2 Safety and risk management issues over construction period

As mentioned above, new safety documents have been introduced by the Directive 92/57. The document's content has been defined in a broad sense with a view to cover the whole construction industry. But what shall be the adaptable attribute of the safety and health plan for a tunneling project that makes the document useful? Whereas the need for a risk management procedure is not clear defined in the directive and the most national enforcements, it seems to be the most effective tool to prevent injuries and professional diseases during the construction of the tunnels. There are some niceties in describing the relationship of safety and health plan and risk management.

- "Safety and health plan is a tool for working out risk management". Safety and health plan is an instrument based on legislative provisions regarding liability, appositeness, elaboration, approval and focusing mainly on consequences to life, health and partly environmental harm. Stand to reason considering safety and health plan as a subset of risk management. New risks, ineffective overall risk management and side effects may arise when safety and health plans and risk management are considered as identical.

- "Safety and health plan is a tool to prevent injuries and professional diseases during the construction of the tunnels by using risk management procedures and techniques or incorporating parts of the independently carried out risk management"

The latter approach is the most rational one to avoid new risks, ineffective overall risk management and side effects. However, risk management methods and techniques should be used when elaborating safety and health plan. Given that the existing available statistical data on occupational injuries and illnesses for tunneling industry, at present, cannot be considered as sufficient to support a full quantitative risk analysis, the most in use methods to work out a risk analysis are qualitative or semi- quantitative methods. As long as disregard of legal statements is not registered as a determining factor of the hazard occurrence, improvement of statistical data cannot be attained.

As discussed in a previous chapter, hazard identification may be the most critical step in risk management procedure. Tunnels, in general, are:

- Complicated works challenging sophisticated engineering and organizational issues.

- Structures constructed in a natural material with characteristics and parameters of high uncertainty so that the risk of geotechnical and structural collapses may be particularly high.

- Confined places with huge transportation activity and use of bulky machinery.

- Significant energy is required for decomposition and breaking of geomaterial.

Based on the above key characteristics of tunnels, a big variety of hazards may arise, qualifying tunneling as a risky activity calling for high standard mitigation measures. Hazards categorization is attempted below:

- Unanticipated ground conditions

- Incorrect realization of design provisions during construction

- Unpredicted ground behavior

- Gas (natural or related to human activities) intrusions

- Water ingress

- Use of explosives

- Extensive utilization of machinery and transportation activity

- Use of compressed air and electrical equipment in humid and dusty environment

- Unfavorable environmental conditions inside tunnel with the presence of dust, diesel combustion products, explosion fumes, vapor

- Exposure to high level of noise and vibrations

- Working at height, on slippery surfaces and lifting heavy loads

- Difficult ergonomic conditions

- Use of chemical substances

- Contaminated ground

- Structural collapse

- Fire in tunnel

- Inadequate desk study

- Failure of systems (e.g. ventilation, lighting, pumps, power supply)

- Overhead suspended objects

- Failure of access roads

Most of these safety items are well treated with the safety and health plan and file documents. A particular issue is the compatibility of the new legislative framework with the existing traditional standards, rules and regulations. For jointing new legislative framework with the existing traditional standards, rules and regulations, there is an urgency to dispose new generation standards and other regulative texts. In Greece, elaboration of new generation standards [35] is currently underway. Finally the achieved level of safety will be in close relation to the quality of regulatory regime and the effectiveness of monitoring the observance of law statements.

6   Geotechnical and structural safety issues

Numerical analysis is the ultimate tool that can deal with complicated problems, like extensive stress redistribution at the stage of excavation and installation of primary support. Considering as bearing element the system of ground, primary support and corresponding interfaces, it is reasonable to expect an overall safety factor for the whole system. But this is not the case. Since the overall safety factor cannot be defined, separation of the system components has become the accustomed practice, however resulting in overemphasizing the primary support and lessening the role of the ground. The major problem when managing safety margins will be the dual role of ground as load imposing and load bearing element.

6.1   Structural design of primary (shotcrete) lining

The individual loads acting on the primary support are:

- Loads related to stress redistribution at the stage of excavation and installation of primary support. Ground behavior, rate and time of load enforcement and construction sequence have to be taken into account.

- Loads related to time and/or stress dependend behavior of the ground up to the activation of final lining.

- Hydraulic loads (rarely).

- Loads related to construction activities e.g. passage of machinery, grouting, excavation of adjacent tunnel (in case of double tube tunnels), etc.

- Loads related to overhead surface activities or constructions.

The utilization of numerical analysis (focusing mainly on finite elements or finite differences) in conventional design operations should be realized by using characteristic values for actions F_k and ground parameters X_k, that is all partial safety factors must be considered as unit. Thereby numerical analysis will result in deriving combinations of moment and thrust with characteristic values. The next step shall be dimensioning of primary support elements. The best working codes for dimensioning shotcrete lining are DIN 1045 and Eurocode 2. Wu [36] noted on the use of these codes: "The global safety factor in DIN 1045 is not suitable for dimensioning shotcrete lining, since the safety margins for the internal and external forces are apparently different for underground works. In this respect, the code EC 2 presents an improvement. However, the major uncertainty due to inhomogeneity of steel and concrete is negligible as compared with the scatter of the ground properties". Major shortcomings for the use or these codes can be summarized:

- In DIN 1045 design values for actions come up by multiplying characteristic values by a global safety factor (1.75 for ductile failure up to 2.1 for brittle failure) while safety factors for materials are unit. A global safety factor does not make any difference on the failure attributed to bending moments or thrusts. In general bending moments may be proven of less importance since shotcrete creep and reduced bending stiffness (which has to be treated by non linear analysis) may reduce their values while anticipated ground movements may provide an effective alert.

- EC 2 uses partial safety factors for deriving design values, both for actions and materials, without making as well distinction of bending moments and thrusts. Non-rational management of uncertainty by ascribing specific numerical values to partial safety factors can be considered as the main problem for tunneling applications.

Wu [36], based on the above concepts, proposed his own scheme of safety factors, trying to raise the aforesaid limitations of these codes. The basic idea is to ascribe different safety factors for bending moments and thrusts and hold unit safety factor for materials.

In fact, the EC 2 and DIN 1045 codes have been developed and widely used for the structural design of above ground structures as obligatory and constraining norms. Such an obligatory and constraining norm for design of primary lining is missing, so that there are the following alternatives:

- Uncritical adopting of above ground structures norm. To a great extent this choice reflects the pressure for a normative umbrella, as long as it is considered as one-way. The lack of a rational and counterbalanced risk-based design concept for primary lining may have effects in sub-optimized design from an economical aspect.

- Dimensioning of primary tunnel lining can be done using safety factors chosen by engineering judgement, made for each individual tunnel. Regarding engineering judgement, this will take into account the perceivable relative uncertainty and in general the outcomes of a risk analysis too.

- Dimensioning of primary tunnel lining can be done using partial safety factors chosen on the basis of existing codes. It is beyond the scope of the on hand presentation to examine closely this special item. As starting point, CIRIA Report 63 [37] can be considered. Research, already accumulated knowledge (quite suggestively Diamantidis [38]) and moreover formalized statistical data can form the basis for calibration.

The constitutive law of shotcrete has to be considered as of high importance and to be incorporated into analysis and simulation procedures, directly or indirectly, in a proper way.

6.2 Structural design of final lining

Tunnel final lining has to withstand the following actions:

- Permanent actions
    1. Self weight of the lining
    2. Any construction permanently resting on bearing structure (e.g. permanently suspended equipment, invert filling)
    3. Geostatic actions from the surrounding ground
    4. Hydrostatic pressure provided the tunnel lies under the water table level and release arrangement is not included
    5. Internal hydrostatic pressure, in case of hydraulic pressure tunnels
    6. Loads from above ground structures or adjacent structures (existing or foreseen)

- Variable actions
    7. Live loads corresponding to standard vehicles (as many as the traffic lanes, in case of invert)

8. Shrinkage

9. Temperature variation

10. Loads related to construction activities e.g. passage of machinery, grouting, excavation of adjacent tunnel (in case of double tube tunnels), etc.

11. Loads related with overhead surface activities

12. Air pressure changes due to train velocity (for railway tunnels)

- Accidental actions

13. Explosion

14. Fire

15. Impact

16. Hydrostatic pressure, due to impermanent obstruction of pressure release system (usually considered as uniformly distributed load 50 kPa)

17. Hydraulic transient pressure, for hydraulic pressure tunnels

18. Earthquake

Design of tunnel final lining is carried out to a great extent using codes for above ground structures, not predestinated for underground works. ITA Working Group on General Approaches in the Design of Tunnels [39] noted "National codes for concrete or steel structures may not always be appropriate for the design of tunnels and the supporting elements. Computational safety evaluations should always be complemented by overall safety considerations and risk assessments employing critical engineering judgements...". The aforesaid remarks for primary support are valid as for final lining too, although lesser as interaction here may be of minor importance after achievement of equilibrium and stabilization with primary support.

Some issues related to design of final lining often call for a more rational approach. For example different tunnelling schools deal with the contribution of primary support in long term support requirements in a very divergent way. It is a common practice to ignore any contribution of primary support in long term support requirements. The rationale can be briefly summarized:

The contribution of shotcrete is ignored on account of reduced durability and creep behaviour. Factors governing durability of shotcrete:

- Exposure conditions

- Materials and construction methods

- Stress and deformation history especially for the young shotcrete.

The long term deterioration of shotcrete associated with alkali based accelerators is no longer an issue with the advancements in alkali free accelerators. Moreover applying contemporary specifications the quality of the shotcrete can be at least comparable to that of conventional concrete, therefore there are no reasons excluding a priori shotcrete from long term support elements.

The decision of including or not shotcrete in long term support elements calls for deep understanding of all the above three factors and some criteria for exposure conditions (that can be taken from concrete codes) and stress and deformation history. Based on field observations and research, a phenomenological approach can be formulated: "Shotcrete linings do not develop any remarkable damage when the displacements are less than corresponding to a critical strain 0.8-1%. Long term strength and behaviour does not influence unfavourably when the imposed stress does not exceed 70% of shotcrete characteristic strength" [40]. This type of statement enhanced with more field observations and research data can aim and support decision for the role of shotcrete. It has to be noted that even considering long term role of shotcrete as a Mohr-Coulomb material ($c = 0, \varphi = 30° and E = 10\ GPa$) in some cases may result in remarkable savings [41].

The contribution of anchors is ignored on account of reduced durability and creep behaviour of anchor elements and ground too. In any event it is not wise to exclude a priori anchors without evaluating the aforesaid factors.

References

[1] Beck Ulrich (1992) "Risk society: Towards a new modernity" Sage Publ., London.

[2] European Commission (1999): Durability and the Construction Products Directive, Guidance Paper F, European Commission, DG III, Brussels, Belgium.

[3] European Organization for Technical Approval (1999): Assumption of working life of construction products an guidelines for European Technical Approval, European Technical Approvals and harmonized standards, EOTA Guidance document 002, European Organization for Technical Approval, Brussels, Belgium. December.

[4] European Organization for Technical Approval (1999b): Assessment of working life of products, EOTA Guidance document 003, European Organization for Technical Approval, Brussels, Belgium. December.

[5] International Organization for Standardization (2000): ISO 15686-1 Buildings and constructed assets – Service life planning – Part 1: General principles. International Organization for Standardization, Geneva

[6] International Organization for Standardization (2001): ISO 15686-2 Buildings and constructed assets – Service life planning – Part 2: Service life Prediction Procedures, International Organization for Standardization, Geneva

[7] Gehlen G. & Schiessl P. (1999): Probability – Based durability design for the Western Scheldt tunnel. In Structural Concrete Journal of the fib. Vol. P1, No2, pp1-7

[8] Breitenbuchner R., Gehlen C., Schiessl P., Van den Hoonard J., Siemes T. (1999): Service life design of the Western Scheldt tunnel, proc. 8DBMC, eds M.A. Lacasse & D.J. Vanier, NRC Research Press, Ottawa, pp. 3-15

[9] Gehlen G. & Pabsch G. (2003): Durability assessment of a railway tunnel. (Re)Claiming the Underground Space, Saveur (ed.), Swets & Zeitlinger, Lisse, ISBN 90 5809 542 8, pp. 125-130

[10] Guidelines for Road Design- Issue 8: Road Tunnels, Civil Engineer Works (2003) in Greek, YPEXODE DMEO

[11] Fischhoff B., Lichtenstein S., Slovic P., Derby S.L., Keeney R.L. Acceptable Risk. Cambridge University Press , New York, 1981

[12] Meacham B.J. (2001) Understanding risk, Proceedings of the ABCB Conference Building Tomorrow's Future – International and National Partners, from web site: www.arup.com/DOWNLOADBANK/download254.pdf

[13] Eskesen S.D, Tengborg P, Kampmann J, Veicherts T.H. (2004) Guidelines for tunneling risk management: International Tunnelling Association, Working Group No 2. Tunnelling and Underground Space Technology 19 (2004) pp. 217-237

[14] S.N. Jonkman, J.K. Vrijling, P.H.A.J.M. van Gelder B. Arends, (2003) Evaluation of tunnel safety and cost effectiveness of measures. In Safety and Reliability - Bendford & van Gelder (eds) pp. 863-871.

[15] S.N. Jonkman, P.H.A.J.M. van Gelder, J.K. Vrijling, (2003). An overview of quantitative risk measures for loss of life and economic damage. Journal of Hazardous Materials A99 (2003), pp. 1-30

[16] J.K. Vrijling, W. van Hengel, R.J. Houben (1998) Acceptable risk as a basis for design, Reliability Engineering and System Safety 59, pp. 141-150

[17] H. Bohnenblust, Risk- based decision making in the transportation sector, In R.E. Jorissen, P.J.M. Stallen (Eds), Quantified Societal Risk and Policy Making, Kluewer Academic Publishers (1998)

[18] Railway Safety. Profile of safety risk on the UK mainline railway. Report No SP-SRK-3.1.3.11. Issue 3 (February 2003)

[19] Health and Safety Executive (2001). Reducing risk protecting people. HSE's Decision making process HMSO.

[20] Rainer Bell, Thomas Glade, Marco Danscheid, (2004) Challenges in defining acceptable risk levels. In Coping with risks due to natural hazards in the 21st century – Risk 21.

[21] Institute of Chemical Engineers (1985) Nomenclature for hazard and risk assessment in the process industries ISBN 85 295184 1.

[22] ICOLD 2002. Risk Assessment in Dam Safety Management: A Reconnaissance of Benefits, Methods and Current Applications ICOLD Bulletin, Draft, August

[23] Robert Melchers (2002), Safety and risk in structural engineering. In Progress in Structural Engineering and Materials. Volume 4 Issue 2 pp. 193-202

[24] Ball D. J., Floyd P.J. (1998): Societal risks. Final report. Available from Risk Assessment Policy Unit, HSE

[25] Directive 2004/54/EC of the European Parliament and of the Council on the minimum safety requirements for tunnels on the Trans - European road network, Official Journal of EU L167/04-30-2004

[26] Guidelines for Road Design- Issue 9: Road Tunnels, Electromechanical Works (2003) in Greek, YPEXODE DMEO

[27] EN CR 14380: Lighting applications – Tunnel Lighting

[28] ANSI/IESNA RP-22-96: American National Standard Practice for Tunnel Lighting

[29] Recommendations of the group of experts on safety in road tunnels, Final Report December 2001, United Nations – Economic and Social Council – Economic Commission for Europe

[30] Structural Safety 2000-01, Thirteenth Report of the Standing Committee on Structural Safety (2001), The Institution of Structural Engineers ISBN 0 901297 16 X

[31] The collapse of NATM tunnels at Heathrow Airport (2000). Health and Safety Executive, HMSO.

[32] European Union Directive 92/57/EEC (1992) Iimplementation of minimum safety and health requirements at temporary or mobile construction sites, Official Journal of the European Communities No L 245.

[33] European Union Directive 89/391/EEC (1989) Introduction of measures to encourage improvements in the safety and health of workers at work, Official Journal of the European Communities No L 183.

[34] Tietz S,B. (1998), Risk analysis – uses and abuses, The Structural Engineer, 76, pp. 395-401

[35] Provisional National Technical Specifications (2004-2005) in Greek, from web site: www.iok.gr

[36] Wu W., Rooney P.O. (2001) The role of numerical analysis in tunnel design, in Euro-summerschool Inssbruck 2001, Eds D. Kolymbas, ISBN 3-89722-873-4, pp. 87-168

[37] Rationalization of Safety & Serviceability Factors in Structural Codes, (1977) CIRIA Report 63, London.

[38] Diamantidis, D. and Bernard, S. (2004), Reliability-based resistance design of FRS tunnel linings, Proceedings 2nd International Conference on Engineering Developments in Shotcrete, Cairns, Australia, October 4-6.

[39] ITA Working Group on General Approaches in the Design of Tunnels, (1988), Guidelines for the Design of Tunnels, Report edited by Heinz Duddeck in Tunnelling and Underground Space Technology Vol 3, No 3 pp. 237-249

[40] Bakoyannis Y, (2004) Actions, methods and models for analyzing tunnel final lining, in Analysis and Dimensioning of Tunnel Final Lining, Eds Tassios Th., Athens 2004.

[41] Watson P.C., Warren C., Hurt J.C., Eddie C., (2001), The design of the North Downs Tunnel, Underground Construction 2001 ISBN 0 7079 7033 4 pp. 63-77

Engineering decisions based on hazard assessment

Dieter Fellner

Electrowatt Infra Ltd., Hardturmstrasse 161, CH-8037 Zurich, Switzerland

Abstract: For designing and dimensioning tunnels geology has to be transferred into technical decisions as choice of excavation method or selection of support types. If one compares international projects and their methodological approaches one can conclude that rockmass classifications are still of outstanding importance in this transfer. This is astonishing as hazard based approaches are common practice in geotechnical planning and designing of construction works close to surface. In our opinion this approach should be applied also for tunnels independent of their depth. In this paper we compare different methodological approaches and their "cooking recipes". Some of their weak points and theoretical deficiencies are highlighted. We outline how tunnels can be planned on the basis of hazard identification → assessment → analysis and how counter-measures might be planned systematically. Thus we support the Swiss attitude expressed well in the outstanding tunnel project Gotthard Basetunnel, not to use "black-box-classifications" but hazard assessment as the basis for engineering decisions.

Keywords: hazard assessment, deep seated tunnels, rock mass properties, GSI, rock mass classifications, flysch type rock, selection of TBM

1 Introduction

For designing and dimensioning tunnels geology has to be transferred into technical decisions. For the results of that transfer it is often the designing engineer who is responsible. For a good result a close collaboration between geologists, geotechnical engineers and designing engineers is essential. The two most important / major decisions are:

- The decision for the apt excavation method often including "TBM-yes-or-better not"-statements

- The decision for an adequate bundle of support types covering the expected range of rock mass behaviour

For both principal decisions there is often a necessity to declare already in very early planning stages, whether TBM excavation is principally feasible or not. The choice of typical support types and an estimated distribution along

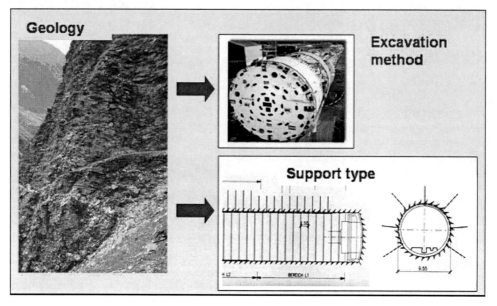

Figure 1: Transfer geology → principal engineering decisions

the tunnel predetermine to great extent the initially estimated costs and time schedule of a project. Thus very important decisions have to be taken at a moment, when the geological uncertainty is high and refined calculations or investigations are still ahead.

If one compares international best practice for various tunnel projects worldwide concerning their methodological approach one tends to end up with the following findings:

- There is an extensive use of rock mass classifications worldwide

- Simplified calculations based on the GSI-approach (GSI = geological strength index [1]) are very popular

- Numerical calculations are in most cases based on the FE- or FD-method with smeared average rock mass properties not distinguishing between intact rock strength and joint strength.

In general there are three possibilities to arrive at the two principal decisions (excavation method / bundle of support types):

1. Taking advantage of experience in comparable tunnel projects. Often similar geological-hydrogeological conditions are required for that approach, as it is often the case in urban areas when planning another metro line additional to already existing ones.

2. By following / trusting empirical rules as provided by rock mass classification recipes. As rock mass classification is derived by statistical analyses of numerous tunnel projects one could state that "classifications are nothing but statistically extended experience"

3. Numerical calculations ("fully engineered solution").

In this paper we support the attitude/approach to base decisions, calculations and the whole geotechnical-related engineering design from early to late project stages fully on hazard assessment. That means choosing relevant hazard scenarios in early project stages based on rough geological information, carrying out calculations for predefined hazards during hazard analysis or making comparison with projects in similar geology to support the hazard assessment.

As international best practice obliges the consulting engineers to produce reports in a way that independent checking engineers can follow in detail and approve the reports, solutions 1 and 2 are generally only accepted in early planning stages, and detailed analysis FE calculations / approach 3 is deemed necessary in most cases. For deep construction pits in urban areas within central Europe national standards demand quite a lot of calculations to be carried out. E.g. calculations/checking for

- hydraulic base failure,

- horizontal static equilibrium

- vertical static equilibrium

- admissible settlements

- adequate length/capacity of support of pre-stressed anchors or braces.

Thus, behind all the calculations required for e.g. a construction pit there are hazard scenarios, that have to be avoided by adding required safety factors. In simple words, one could summarize: hazard * FOS = design hazard. This simple example highlights the fact that ordinary geotechnical consultancy close to surface is fully based on the hazard approach. For deeper seated tunnels there is also a trend e.g. in Switzerland / Austria to take the hazard scenario as a basis for planning and deciding already in early planning stages and not to use classifications at all.

Comment: Concerning the use of rock mass classifications it is interesting that there are big differences in Europe. The Scandinavian countries tend to use Q-method extensively. In an actual Island Hydropower Projects no classifications have been used although belonging to the Scandinavian Northern part of Europe. In Italy the RMR is very popular. In Switzerland for the Gotthard Basetunnel and in Austria for the Brenner Base Tunnel and the Wienerwaldtunnel — just to mention a some outstanding projects — no classifications have been used at all or just additionally.

To highlight the methodological differences for planning a tunnel some selected tunnel projects are mentioned below.

2   "Recipes" and examples

Example 1 / Tunnels Krimnou & Dyo Koryphon (Egnatia Odos, Greece) (Fig. 2)

The two tunnels are part of a series of D&B tunnels in mainly flysch type rock with a maximum overburden of 175 m. The rocks have been grouped according to their relative percentage of sandstones/siltstones in categories A-E. These categories have been transferred into units along the longitudinal profiles of the tunnels. Various classifications and empirical recipes (RMR, Q, GSI) have been applied. The rock mass properties have been derived empirically and used as smeared average values in FE calculations. For arriving at design rock mass parameters for each category, GSI-values have been derived via RMR (GSI=RMR89 − 5) [6] and average uniaxial compressive strengths have been estimated according to the relative percentage of sandstones. For additional wedge calculations a different set of joint design parameters has been proposed.

Example 2 / Kangle & Shanggongshan (Kunming Water Supply Project, China) (Fig. 2)

The geology consists predominantly of basal sandstone/siltstone alternations (flysch type rock) and overlying dolomitic rocks. The rock mass properties have been derived empirically [9] and used as smeared average values in convergence-confinement calculations. The GSI recipe for flysch type rocks according to Hoek & Marinos [6] served as a basis for deriving rock mass properties. Again, average uniaxial compressive strengths have been estimated according to the relative percentage of sandstones. By use of convergence-confinement calculations six rock mass behaviour categories have been proposed (a-f). Whether a rock falls in a certain behaviour category or not, is defined by rules that can be derived by simple convergence-confinement calculations: the "strength to stress ratio", the extent of the plastic zone and the squeezing potential. For details we refer to [10].

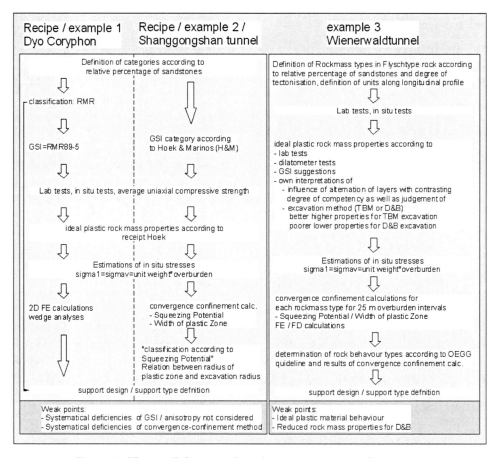

Figure 2: "Recipes" for geotechnical prognosis case studies 1+2+3

Example 3 / Wienerwaldtunnel (Austria) (Fig. 2)

The geology again consists mainly of "flysch type rock" with variable amounts of sandstones. The geotechnical approach has been based principally on the recommendations of the Geomechanical Guideline OEGG [8] and Fig. 2. 17 different rock types have been described. Different rock mass properties have been suggested for TBM and D&B excavation separately, which is astonishing. The rock mass properties have been derived empirically partly based on the GSI approach, partly based on judgment of lab and in situ tests. The properties have been used as smeared average values in convergence-confinement as well as in FE/FD calculations.

Short résumé

- All three mentioned projects deal with anisotropic rock

- All mentioned projects took advantage of GSI-derived rock mass properties, with the difference that in the Wienerwald tunnel project the GSI-derived properties have not been taken directly but were somehow adopted.

- As the GSI approach merely delivers one value for cohesive strength, friction and YOUNG's modulus (modulus of elasticity) all projects were calculated with ideal plastic rock mass behaviour (convergence-confinement or FE/FD calculations).

- Whereas the Wienerwaldtunnel is mainly based on rock mass behaviour types (treated as "hazard approach") the examples in Greece and China rely to a much greater extent on empirical rules (recipes).

This short comparison indicates also the trends in Switzerland and Austria to use as little as possible recipes or — when using — critically judged. The reasons for a critical attitude are manifold: tradition, negative experience, dislike of black-box-approaches, lots of tunnels facilitating the use of experience,...

Additionally the following systematic deficiencies can be highlighted:

- The GSI approach in the present version is not suited for anisotropic rock mass. Assigning a low value for uniaxial compressive strength is from practical point of view the only possibility to reduce the rockmass strength if deemed necessary.

- Of course, if large deviations from ideal-plastic behaviour are expected (strain softening) one shouldn't expect GSI proposals to be correct.

- The GSI approach and classifications in general have deficiencies, when dealing with fault zones of limited width (< 30 m), as none of the recipes proposed so far take the decisive factor "fault width" into account. From experience a well trained foreman often can judge from the fault width whether a fault zone will not even be recognized from engineering point of view or cause significant problems.

- Concerning fault material one has to mention also that judgements concerning intact rock strength are not possible for crushed or heavily tectonized rock.

- Without going into further details, one should bear in mind that e.g. the RMR classification doesn't take the overburden into account, which poses a big question mark when using RMR for deeper seated tunnels.

To cope with those deficiencies and before — hopefully soon — updated GSI or other new recipes will allow rock anisotropy to be incorporated to some extent in parameter determination and calculations, we recommend the following numerical solutions for overcoming the limitations mentioned above

- The use of anisotropic constitutive material behaviours distinguishing between matrix and joint properties, if anisotropy is expected to be of increased relevance

- To model the wall effects of adjacent more competent rock on fault zones by using axisymmetric or 3D-models

- To study the influence of non ideal-plastic behaviour, e.g. strain softening, on deformations by using strain softening or non linear constitutive equations.

3 Some comments on rock mass properties and deep seated tunnels

As mentioned before, the GSI approach delivers only one value for rock mass friction, cohesive strength and YOUNG's modulus (modulus of elasticity) which means no strain softening and no anisotropy is taken into account. There are further important factors that might play a major role especially for deeper seated tunnels.

- Stress dependent softening properties; higher confining pressures in the vicinity of the tunnel lead to less pronounced strain softening in comparison with tunnels close to the surface.

- Stress dependent deformation properties: The YOUNG's modulus (modulus of elasticity) is often increasing with depth; this tendency has been observed by comparing lab results of Gotthard Basetunnel taken close to the surface and at great depth. Lab tests on marly schists for a nuclear repository in Switzerland showed the same trend.

- Stress dependent permeabilities: Permeability is often decreasing with increasing depth by several orders of magnitude. This is the case especially in moderately weak to soft rock and less pronounced in very hard granite type rock.

Although these trends are significant, they are difficult to be taken numerically into account because stress dependent constitutive laws are rare in commercial available software packages. Looking at the fact that a stress-dependent YOUNG's modulus (modulus of elasticity) can reduce the radial deformations by e.g. 10 cm and more, it gets evident, that these factors might be of great importance for deep seated tunnels, especially in bad rock conditions. Fig. 3 shows the depth-dependent increase of the YOUNG's modulus (modulus of elasticity) as well as decreases of permeability and strain softening. Similar trends can also be predicted in the vicinity of the tunnels where stresses are increased at the outer rim of the plastic zone and decreased at the excavation surface.

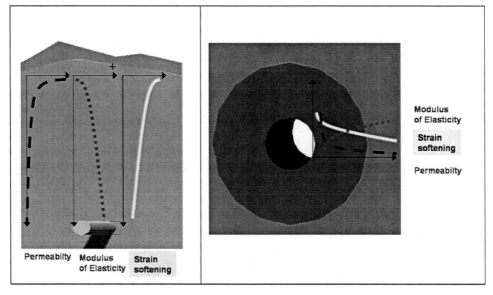

Figure 3: Stress-dependency of rock mass properties for deep seated tunnels — illustration of principal trends

Comment on the reduction of values for friction and cohesive strength:

For slope stability issues the standards suggest a reduction of internal friction and cohesive strength to take uncertainties of the rock mass properties into account. For deep seated tunnels such general reduction of strength properties is not recommended as it can happen that a still feasible tunnel turns into non feasible. In other words: for deep seated tunnels the influence of a largely reduced cohesive strength / friction angle on deformations is highly nonlinear and the proposed reduction factors for e.g. slope stability issues are not applicable.

4   How does the hazard approach work?

The principle of hazard approaches is simple. Based on the geological data and subdivided geological and lithological units one defines principal hazards (hazard identification). The hazards are weighed and a subdivision is made between dominant and subordinate hazards (hazard assessment). Further on, hazards are analyzed to get the intensity and probability right (hazard analysis). Then, all hazards are comprised in a project-specific hazard register. The counter part of that hazard register is an "action register" including all measures to counteract the hazards. Of course, the various support types are (or should be) the principal answers on the most frequent combinations of various hazards. To cope with rare hazards, the so-called additional measures — as e.g. pre-grouting — are described in the action register. The additional measures extend the applicability of principal support types. One could say, that everything (hazards and measures) finally determine a big matrix. Various measures are defined to counteract hazards according to their intensity. The matrix thus also shows which support types are apt for counteracting various hazards depending on their intensity. The principle of that "matrix" is shown in Fig.

hazard register								
H 1			H 2					
low intensity	moderate intensity	high intensity	low intensity	moderate intensity	high intensity			
x	x		x			AST1	support type	action register ("measures")
		x		x		AST2		
					x	AST3		
		x		x		B1	additional measures	
					x	B2		
					(x)	B3		

Figure 4: Principle of final decision matrix in hazard approach

Example - hazard approach for Gotthard Basetunnel

The 57 km long Gotthard Basetunnel — which is presently under construction — will be the longest railway tunnel in the world. During planning the central part (lot Sedrun), the approach based on hazard scenarios has been fully applied. The depth of the investigations and analyses has been successively increased during various planning stages. Concerning hazards this means: the hazards have first been judged mainly qualitatively and analyzed in great detail in late project stages. The procedure has been already described in various publications and will only be outlined below. For further details we refer to published papers. [1, 4, 5]

In early project stages the various rocks have been comprised in so-called rock types. Geotechnical units and fault zones have been defined along the longitudinal profile of the tunnel. These units also served as a basis for all following descriptions including judgement of hazards and their intensities. During early planning stages the principal hazard scenarios with 5 intensity classes, described for each geotechnical unit, served as the basis for engineering design and associated cost prognosis. The principal geotechnical hazard types included: plastic deformations, wedges, unravelling, rock burst and spalling. Additionally hydrogeological hazards have been described. The intensity classes 0 to 4 have been defined as follows: 0=no or very low intensity, 1=low, 2=moderate, 3=high, 4=extreme intensity. During detailed design the hazards have been widely extended. Especially — as heavy squeezing rock conditions are expected in that lot — the principal hazard "plastic deformations" has been extended by various subtypes as e.g.:

- even extrusion of the tunnel face when entering weak zone

- uneven extrusion of tunnel face when intersecting inclined fault zone

- uneven plastic deformations due to rock anisotropy

- delayed plastic deformations due to pore water effects or creep

- extrusion of material between yielding steel arches.

Due to the great amount of subtypes of principal hazards the engineers decided to elaborate a database application to organize hazard scenarios and counter-measures and to provide a decision aid for site engineers. For including financial aspects the database application has been extended to clarify which measure is cheaper in case that several measures are possible to counteract a certain hazard scenario.

Numerical tools have been programmed to deliver fast deformation prognoses based on actual updated geological findings in pre-probing [3].

5   Geotechnical hazards and calculations

Thinking in hazard scenarios has the following general effects on calculations when designing tunnels. That approach calls for:

- Some calculations already in an early project stage during investigation of the general rock mass behaviour. Most calculations are done by engineers for demonstrating that the factors of safety of the suggested support are acceptable. To investigate e.g. unravelling in a UDEC model, realistic geological joint data are of primary interest and it would help to have more "calculating geologists".

- More realistic models for the rock mass behaviour.

- An increased number of discrete element calculations.

The hazard approach and the necessity of realistic models make deficiencies and limitations evident. So during investigation of hazard scenarios one soon recognizes that not all hazard scenarios can be analyzed numerically due to an excessive number of detailed scenarios or other reasons.

Example 1: The intersection of a vertical fault with a tunnel is easily modelled. To model the intersection of an inclined fault intersecting the tunnel axis at an oblique angle is very time consuming and it is often not possible for consulting companies to carry out full numerical analyses. Time and cost aspects very soon lead to the following admissible reaction: Calculate a simple case and extrapolate the results on more complicated cases.

Example 2: Rock burst hazard: there are several approaches described to estimate the risk for rock burst hazard (m=0 method described by Kaiser et al. [14]), but one generally has to admit that the numerical results are not very reliable.

Example 3: Water saturated sandy dolomite: Recent excavation of a TBM in Southern China showed water and sand ingress in the TBM when entering such fault zone. This hazard cannot be modelled by normal commercial codes available.

Example 4: Wedges and stresses: The well known program UNWEDGE [6] allows to take boundary stresses for simple wedge geometries (tetraeder) into account; but not for more complicated wedge-geometries. Also fracture generation is not included.

These few examples mentioned above should demonstrate that numerical limitations are encountered very soon and that the attitude to try to calculate everything is sometimes illusionary. As many hazard scenarios can be encountered with conceptional solutions, precise numerical analyses are not always necessary. The following example should demonstrate a combination of numerical calculations and conceptional solutions.

6   Example for hazard approach in execution stage

The following example is a simplified theoretical example demonstrating the kind of hazard approach during construction of a deep seated tunnel, when entering a major problem zone. The tunnel was systematically accompanied by pre-probing. They showed a 30 m wide fault zone ahead of tunnel face parallel to schistosity; the strike of the schistosity was normal to the tunnel axis, the dip was 70° in direction of tunnel excavation. The excavations showed that the adjacent more competent gneiss-type rocks can be expected to show deformations in the range of < 1 cm, although heavily fractured. Unravelling is here the main hazard scenario. Due to an overburden of 800 m the in-situ vertical stresses can be assumed to be in the order of 21 MPa. The horizontal stresses are estimated to be in the same order of magnitude ($K_0 = 1$ assumption). Sequential numerical calculations (analogue to section 4) carried out for different parameter sets for the described case but with the simplified assumption of a strictly vertical orientation (necessary for axisymetric calculations) gave the following numerical expectations:

- Significantly increased tunnel face deformations when entering fault zone, numerical sliding micrometer showed stepwise increasing deformations up to 15-35 cm.

- Inside the fault zone: 12-30 cm radial deformations.

- Last third of fault zone: significantly lower deformations at face as well as radial.

- General very large plastic zones also in competent rock at both rims of the fault.

Qualitative considerations lead to further expectations as (Fig. 5):

- Uneven radial deformations in the crown when entering fault zone.

- Uneven face deformations when entering fault zone; first predominantly in upper part of tunnel face.

- Significantly increased risk of unravelling in transition zone between adjacent rock / fault zone.

- Eventually long term deformations.

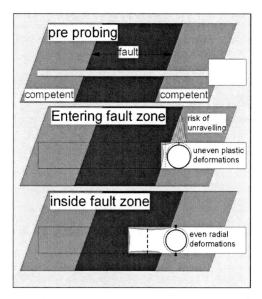

Figure 5: Fault zone - expected rockmass behaviour/ hazards

The hazards are shown schematically in Fig. 5. Countermeasures based on calculations and qualitative considerations can be formulated as follows:

Hazard scenario/expectations	Counter measures
Uneven radial plastic deformations (with moderate to high intensity) when entering fault zone deformations are expected to be increased in crown of tunnel	Radial over-excavation (in tunnel crown) and yielding steel support in combination with systematic radial bolting; 30 cm radial deformations in yielding stage of steel arches, monitoring with 3D optical measuring points, survey sections each 3 m
even plastic deformations with high intensity	
Significant decreasing radial plastic deformations in second half of fault zone	Possible reduction of excavation diameter
Uneven extrusion of tunnel face with high intensity when entering fault zone	Heavy support by long overlapping steel bolts without face plates
At the rim of the fault zones: significantly increased risk for unravelling; intensity high, radial as well as in tunnel face	Short round lengths (1 m), wire mesh behind steel arches, forepoling, sealing of tunnel face with shotcrete
Long term pressure build up and deformations	Geodetic survey, late application of final lining

7   Hazard approach & TBM feasibility / TBM selection

For the Brenner Basetunnel the squeezing potential has been estimated along the chosen alignment and extended for the whole investigation area. This squeezing potential is nothing else but a deformation prognosis for an unsupported tunnel based on convergence-confinement calculations, systematically carried out for different overburdens. The limit of 5 and 7% squeezing potential has been used for judging the feasibility of large diameter TBMs and smaller double shield pilot-TBMs. Thus a specific hazard scenario "squeezing rock" - encountered where rocks with poor properties are under high tectonic and/or overburden stresses - has been taken as principal decision aid for determining the TBM feasibility in an early project stage.

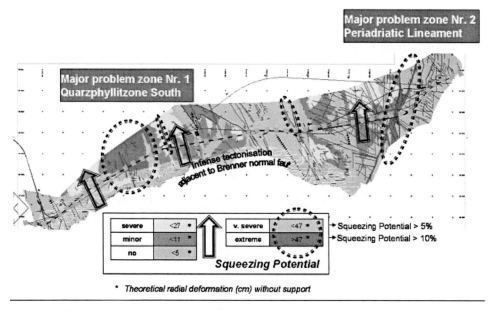

Figure 6: Geotechnical prognosis Brenner Basetunnel: Plastic deformations & areal squeezing potential

Already in early project stages the risk of blocking of the shield of a TBM due to intense plastic deformations when entering weak zone — probably of limited width — should be analyzed. As these kind of judgements are frequent for TBM consultants, Electrowatt is producing their own tools for such purposes by programming FLAC and other ITASCA programmes to produce a fast answer. E.g. by using FLAC axisymetric calculations we determine, what are the deformations encountered at the tunnel face and at a predetermined distance from that and check by that simple means whether blocking of the shield is endangering or not.

During carrying out a pre-feasibility study of the Western Route of the South-to-North water transfer project in China (7 tunnels in line with 260 km total length), the approach used for Brenner Basetunnel (based on the squeezing potential) has been extended in that sense, that also other hazards have been taken into account, less for the purpose of feasibility judgement, but more for the judgement of which TBM should be suggested for what conditions. For long tunnels in hard rock high advance rates are required. They can be achieved either by open TBMs or double shielded TBMs. Single shielded TBM usually cannot compete. A double shielded TBM costs significantly more than an open TBM, so from economic point of view the question is often: are the additional costs for a double shield TBM justified? A hazard based analysis of prevailing geology of a tunnel might serve as a first decision aid for recommendations on selecting an apt TBM. Of course, tables as shown in Fig. 7 have to be adopted to project specific hazards and geology.

Several case histories for TBM problems under various geological conditions indicate that the squeezing potential alone is not sufficient to judge the feasibility of a TBM. A good example for this is the Shanggongshan tunnel (China, Kunming), where EWI is providing site supervision, but was not involved in the geological & geotechnical prognosis. Actual geological problems are due to the fact that the 3.5 m double shield TBM, designed for hard or moderate rock, has to pass through series of oblique fault zones consisting of sandy granular water-saturated dolomites prone to collapse. The fault materials presently encountered at tunnel level are also exposed on the surface and a more rigorous geotechnical hazard analysis concerning the type of TBM and their features would have asked for more refined equipment for systematic pre-grouting ahead of the TBM.

8 Hazard approach & overflow spillway transition tunnel

As last example for demonstrating the importance of hazard approaches, the actual geotechnical planning for overflow spillway tunnels (OVS) of a hydro-electric power plant in NE Turkey is mentioned. The first part of the OVS tunnels are large in size and are actually daylighting caverns. The excavation section for the left bank OVS tunnel is about 18 m high and 16 m wide and the portal is very close to the surface. In critical sections the overburden reaches approximately 15 m and even less, i.e. 9 m when considering the shoulder

		Open TBM			**Double Shield - TBM (DS)**	
typical rock	associated hazard	dist. advantage	light advantage	similar	light advantage	dist. advantage
				criteria for open or double shield TBM — **based on comparison of hazard impact on degree of utilisation**		
weak or heavily jointed rock	squeezing + creep effects	*short shield*		*similar when rock too bad / good enough for DS*		
	gripper instabilities in weak rock					*distinct advantage*
	discs do not turn any longer, flattening effect / discs do not cut any longer, grinding effect			*no difference*		
	unravelling event. stress induced		*in case of sufficient stand up time for open TBM*		*in case of very short stand up time*	
	collapse		*better accessibility*		*DS TBM specially equipped for injection works*	
	asymetric deformation/ buckling		*short shield is less problematic*			
predominantly hard jointed rock	wedges event. stress influenced		*sufficient stand up time for open TBM*		*very short stand up time*	
	rock burst, spalling				*safer*	
esp. internal/rim of fault zones	water blow out, blow out with sand			*similar in case of a not very significant blow out*		
	gas			*no difference*		
very hard abrasive rock	abrasivity			*no difference*		

Figure 7: Hazard analysis and TBM selection (open or double shield TBM)

on the valley side. The geology consists of jointed granodiorite and signs of instabilities have been found, when excavating the benches in this area.

The most important geotechnical hazard scenarios are

- roof collapse,
- pillar instabilities,
- general unravelling.

Simple schematic models — again based on hazard approach — can be used to develop a support strategy as follows.

- With support pressure from inside the tunnel and lateral confining pressure on both sides the simplified model shows that the Terzaghi support pressure is valid and gives stable conditions.

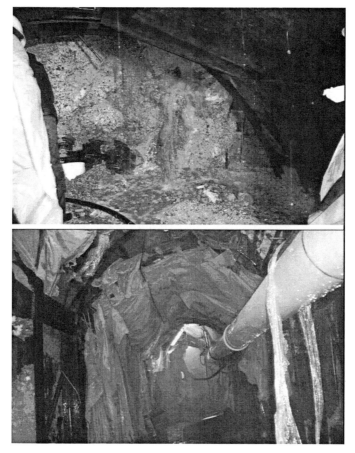

Figure 8: Shanggongshan Tunnel (China, TBM); top: Excavating blocked cutterhead of TBM by sandy granular dolomites in small "cavern"; bottom: By-pass under difficult conditions.

- As one lateral boundary is a free surface, deleting the confinement on that side shows that the Terzaghi support pressure is no valid any longer.

- By means of prestressed anchors above the tunnel crown the "missing lateral confinement" can be re-established. Thus prestressed anchors + support pressure and passive bolting from inside are generally judged as apt means to counteract roof collapse. The subdivision of the huge excavation section in many excavation steps is absolutely necessary. The support pressure from inside is established by reinforced shotcrete and steel arches in the vault.

- Introducing joints in the pillar one can easily model collapse of an unsupported pillar. Again, by introducing passive anchors stable conditions can be obtained. Thus, passive support for the pillar is recommended.

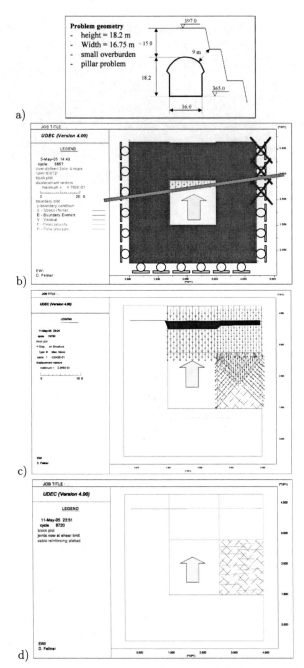

Figure 9: Large span excavation close to surface and principal asociated hazards

a) Problem geometry of OVS tunnel close to surface (original design changed later on);
b)-d) simple models for justification of support strategy: b) prestressed anchors counteract
well hazard "roof collapse"; c) without passive anchoring pillar fails; d) with passive support
in pillar and prestressed anchors above roof model indicates stable conditions

First numerical analyses of this OVS tunnel have been handed over to Electrowatt for checking. The calculations have been carried out with the FE method for different cross sections and for assumed ideal-plastic Mohr Coulomb rock mass behaviour without softening and without joints modelled explicitly. As ideal plastic rock mass behaviour just shows signs of plasticity, but would even without any support demonstrate stable conditions, the FE calculations have been rejected and modelling of strain softening behaviour / modelling of joints have been demanded. We mention this to underline that the question "Which numerical analysis is apt for calculating which hazard scenario?" is a big issue and it is out of scope of this paper to describe it in further detail.

In the meantime these considerations have been agreed by all parties and critical parts of a OVS tunnel have been redesigned.

9   Final comments / suggestions

- Hazard based approaches are common practice for geotechnical planning and design of construction works close to surface. This approach should be applied as well for tunnels independent of their depth.

- Empirical methods like rock mass classifications should only be used in early project stages. For detailed planning / late project stages hazard-based-decisions are absolutely necessary.

- Hazard approaches ask for some calculations already in an early project stage and for more realistic models of rock mass. Going for a hazard approach does not necessarily mean that there are more calculations required.

- Deficiencies of the worldwide used GSI method have been highlighted and a critical perspective were recommended.

- There is a growing tendency to convert the hazard approach to site specific risk management. One should bear in mind that a predefined matrix between hazards and counter measures clarifies only the reactions from engineering point of view. For the geologist the questions: How can I predict zones of weakness, water inflows or wedges, remain the same.

- The way the hazard approach has been presented in this paper is more than an approach: "Thinking in hazard scenarios" is recommended generally as an attitude for planning, designing and supervision. This attitude is confronted with an also very important attitude "thinking in money" and it is not always easy to find the right balance there.

- With the great variety of examples we tried to demonstrate that hazard approaches can be used manifold, e.g. as the basis for detailed planning of a basetunnel, for selecting an appropriate TBM for particular geological conditions or for developing support strategies for an overflow spillway tunnel.

- The question "Which numerical analysis is appropriate for calculating which hazard scenario?" is a big issue and its importance is shortly demonstrated.

References

[1] Fellner, D. (2003): Geotechnische Erfahrungen und Herausforderungen im Abschnitt Sedrun des Gotthard-Basistunnels, *Geotechnical experiences and challenges Lot Sedrun Gotthard Basetunnel*, Felsbau Nr. 3/2003.

[2] Fellner, D. et al. (2003): Brenner-Basistunnel – Geotechnische Prognose und Konzept für den TBM-Einsatz, *Brenner Basetunnel – Geotechnical prognosis and concept for TBM*, Felsbau Nr. 5/2003.

[3] Fellner, D. et al. (2004): Modelling yielding support by programming FLAC2D / FLAC3D, Eurock 2004 & 53rd Geomech. Colloquium, Salzburg

[4] Ehrbar, H. (2004): Alptransit Gotthard – Sedrun – Vortriebskonzept in druckhaften Zonen – Vom Projekt zur Ausführung, Eurock 2004 & 53rd Geomech. Colloquium, Salzburg

[5] Amberg, W. (1999): Konzepte der Ausbruchsicherung für tiefliegende Tunnels, Bauingenieur, Bd. 74

[6] ROCLAB, UNWEDGE, rockengineering: www.rocscience.com

[7] FLAC, UDEC, www.hcitasca.com

[8] OEGG (2001): Richtlinie für die geomechanische Planung von Untertagebauarbeiten mit zyklischem Vortrieb, *Guideline for geomechanical planning of conventional underground works*, Salzburg

[9] Kunming Water supply project, (2001): Bid for the construction, Report on geomechical Characterisation.

[10] Italian Geotechnical Association, (1979): Some Italian experiences on the mechanized characterisation of structural complex formation, Proc. ISRM, Montreaux

[11] Russo G. et al. (1998): A discussion on the concepts of geomechical classes, behaviour categories and technical classes for an underground project. Gallerie e grandi opere sotterranee

[12] Palmstrom, A., RMI: www.rockmass.net

[13] Barton N. (2000): TBM Tunnelling in jointed and faulted rock, www.balkema.nl

[14] Kaiser, P.K. (2002): Brittle failure and tunnelling in highly stressed rock, NDK course, ETH Zurich

[15] Fellner, D., Theiler, A. (2005): Tunnelling through anisotropic rock – experiences from Gotthard-basetunnel lot Sedrun, 54th Geomechanical Colloquium Salzburg, Felsbau 2005/3, in press

Blasting methods in tunneling

Mark Ganster

Austin Powder GmbH, Weißenbach 16, Austria

Abstract: This paper is conceived as a short overview of the newest blasting methods in tunneling. It starts with a description of how to start a tunnel depending on rock quality. Another important topic is the drilling work. Compared to surface blasting we have only one free face in tunnel blasting which means that drilling work should be very precise. The most important thing is to drill a right and exact cut. The cut, the used explosives and the initiation system depend on the tunnel cross section, length of the drill holes, rock to be blasted and on the diameter of the drill holes. The newest technology of on site mixing of emulsion explosives combined with nonelectric detonators allows a very fast and safe charging process. Safety should always be the most important parameter to be respected.

1   Introduction

Eighty to hundred years ago underground construction work was mostly associated with mining activities. In the last decades, however, tunneling has became a more important activity within the field of underground construction work. Tunnels are driven for road and rail, sewer and water supply, hydro-electric power projects, mountain caverns for industrial, recreational and storage purposes and as transport and storage systems for oil and gas. A number of tunnel sites are underneath or close to inhabited buildings or industrial installations. Such conditions require good technical judgment and knowledge of tunnel driving together with careful planning, to ensure safe and satisfactory results. The conventional tunneling technology is progressing rapidly. The performance capability of the construction equipment, in particular that of the drilling equipment, has increased considerably over the last few years. With the introduction of more effective and more accurate hydraulic drilling rigs, with higher capacity of the loading and transportation equipment together with modern and more advanced explosives and detonating systems, loading and blasting techniques have increased the speed of progress in tunneling and reduced construction time and cost. 70 to 80 meter of tunnel advance per week is today quite normal for road and railway tunnels.

2   General

2.1   The tunnel portal and initial approach

When the tunnel portal has been fixed, the area must be carefully cleared of soil and loose rock. The portal is then blasted to give an overburden to the entrance at the least half a tunnel diameter.

The first round is blasted very carefully with minimum borehole depth. In poor quality rock it may even be necessary to start with a smaller pilot tunnel which is subsequently enlarged to the full tunnel dimension.

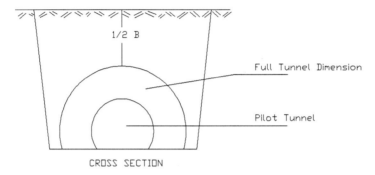

Figure 1: Cross Section of a pilot tunnel which is enlarged to a full tunnel dimension

2.2   Tunnels

It is common practice in rock blasting to drill the holes parallel to a free face to which the explosives in the boreholes can break. In surface blasting work there are at least two free faces - one in front and one on the top surface. In tunneling, however, there is only one free surface, the tunnel face. The advance is perpendicular to the tunnel face. The holes are drilled at right angles to the face so that the face itself can not be used as a free face for breakage. A free face must be generated by a cut which will open a cavity into which the production holes can break. This is achieved by a sequencial blasting technique controlled by the delays of the detonators.

A satisfactory tunnel excavation requires a good cut along the entire length of the production holes. The rock in a tunnel round is more effectively confined than in any other rock blast operation. The physical characteristic of the rock will consequently influence the result of the blast to a greater degree than in other works. The rock in small diameter tunnels is more firmly confined than in tunnels of larger diameter. The effect of this is that the specific consumption of explosive increases with decreasing cross section.

The most important factors which will influence the amount of drilling required are:

- Rock characteristic

- Tunnel cross section

- Borehole depth

- Borehole diameter

- Type of cut used

ANFO has to a great extent replaced cartridged explosives in tunnelling due to price and quick and effective loading techniques with the loading equipment mounted on the drilling rig itself. Under wet conditions cartridged explosives are still being used. The contour holes are normally loaded with column charges or detonating cord. Increased focusing on environmental conditions will most likely increase the use of slurries in tunneling. The use of electric detonators has dominated for years in tunnel driving. For safety reasons non-electric detonating systems are rapidly gaining ground.

3 Drilling

3.1 General

In bench blasting the boreholes are drilled parallel to the free face against which the breakage can take place. In tunneling, however, the only free face available is that of the tunnel face. Drilling parallel to a free face is not possible. To achieve satisfactory blasting results an opening or void - the cut - must therefore be created into which the surrounding rock can be blasted. It is possible with carefully selected initiation systems and with delay detonators suitable for tunnel work to blast the cut and the production holes etc. in one operation.

The creation of a proper cut is a precondition for a satisfactory tunnel blast. A number of different types of cuts have been developed over the years concurrent with the development of the drilling equipment.

The cuts can basically be divided into the following two groups:

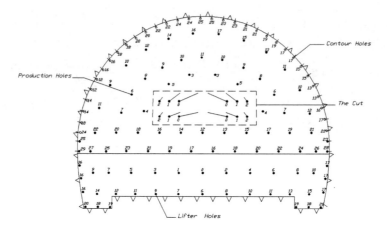

Figure 2: Nomenclature tunnel driving

- V cuts: Wedge cuts and fan cuts

- Parallel cuts: Large diameter and burn cuts

Parallel cut, large diameter, is the most commonly used cut. This type of cut is independent of the crossectional area of the tunnel and tunnel width. It is applicable over the entire cross section of the tunnel and for various borehole diameters. A symmetrical wedge cut is also in use.

3.2   Types of cut

The choice of cut is decided on the basis of the cross sectional area of the tunnel and its width, available drilling equipment, preferred advance and rock conditions. The most commonly used types of cuts are:

- Fan cut

- Wedge cut

- Parallel cut, large diameter

- Burn cut

3.2.1   Fan cut

The principle of the fan cut is breakage / back ripping against the tunnel face. The fan cut is effective with respect to the total length of boreholes being drilled and quantity of explosives being used for each round. The cut gives a comparatively easy breakage, consuming less explosives. The fan cut requires a wide tunnel which limits the advance per blast. A non-symetric drilling makes the cut less effective when applying modern drilling equipment.

3.2.2   Wedge cut

In the wedge cut, the holes are drilled at an angle to the face in a symmetrical formation. The angle of the wedge must not be too small. The burden is increased with increasing angle. The symmetric pattern distributes the drilling work evenly between the drills. The increased burden requires more explosives and more drilling work as compared with the fan cut. The advance depends upon the tunnel width and the wedge cut is consequently most suited in wider tunnels. A wedge cut blast throws the broken rock further out of the tunnel as compared with other types of cuts. This makes it more cumbersome to scale the roof and the face from a position on top of the muckpile and it increases mucking time. The greater throw of fired rock is also more likely to cause damage to ventilation ducting close to the face.

3.2.3   Parallel cut

The parallel cut consists of one or more large diameter unloaded boreholes. All holes are drilled at a right angle to the face and parallel to the tunnel direction. The breakeage is against the opening or void formed by these unloaded holes of diameter 76-150 mm.

This opening is gradually opened up by successive detonation of the adjacent loaded holes and the pulverized rock is blown out of the cut. The parallel cut requires a weak load along its entire length. Packaged explosives of a composition giving gas energy with low gas temperatures are used. ANFO is also to a large extend being used in the parallel cut procedure. The use of explosives with high speed of detonation may easily cause sintering, reduced blow out and an unsuccessful blast. Modern electric/hydraulic drilling equipment will drill up to 152 mm diameter. Several large diameter cut holes will help to secure good breakage along the entire length of the advance. The length of the advance is independent of the cross section of the tunnel. Parallel cuts give in general better breakage and fragmentation of the rock with less throw-out and spreading as compared with the V-cuts.

The advantages of the parallel cut - large diameter are:

- Well suited for drilling with modern drilling equipment

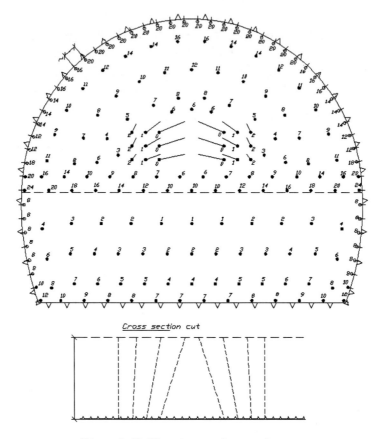

Figure 3: Drilling pattern for a wedge cut

- Well suited for rounds of long length

- The length of the advance is in principle independent of the cross section or width of the tunnel

- Good breakage

- Moderate throwout and spreading of the pile, reduced loading time and scaling of roof, walls and face from a position on top of the pile

- Good fragmentation

It is a precondition of the parallel cut, large diameter, that the rock blasted into the empty boreholes has sufficient room for complete blowout. Based on experience, this requires very accurate drilling and carefully selected delay times and sequence of detonation for the loaded holes adjacent to the cut.

The following conditions must be satisfied:

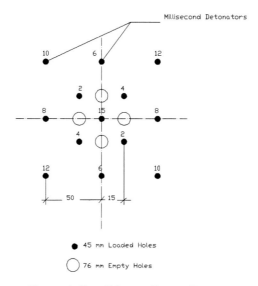

Figure 4: Parallel cut - Large diameter

- Sufficient opening for the rock to be blasted and blown out

- The burden must be in the right proportion to the opening

The distance between the large diameter unloaded cut holes and the adjacent loaded holes is given by the following formula:

$$\text{Burden } B = 1.5 - 2.0 \times d \ [\text{mm}] \tag{1}$$

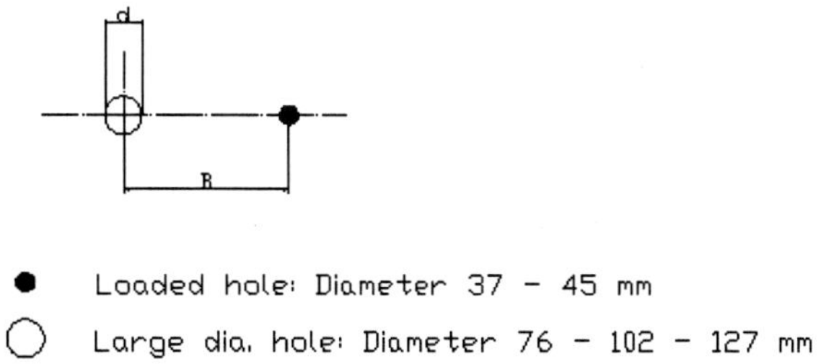

●   Loaded hole: Diameter 37 – 45 mm
○   Large dia. hole: Diameter 76 – 102 – 127 mm

Figure 5: Connection between burden and large diameter hole

The burden B for the other holes of the cut is governed by the width W of the established opening. Based on experience with rock of medium blastability:

$$\text{Burden } B = 0.7 \times W \ [\text{mm}] \tag{2}$$

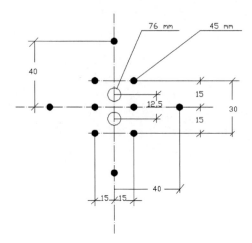

Figure 6: Parallel cut - Large diameter with two empty holes (76 mm)

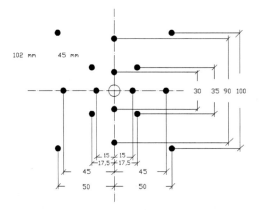

Figure 7: Parallel cut - Large diameter with one empty hole (102 mm)

3.3   Drilling pattern

The drilling pattern design starts by positioning the cut and the contour holes first. Secondly get the production holes adjacent to the contour in place and fix the position of the lifter holes. Finally position the production holes. When fixing the production holes make sure to allow sufficient room for the cut breakage. The offset is between $0.2 - 0.4$ m. The burden in the transition zone between the contour holes and the production holes must be reduced to $0.4 - 0.6$ m.

To ease the drilling and loading operations, it is normal to position the cut $1.0 - 1.5$ m up from the floor and alternately on either side of the centerline to avoid collating a hole into a socket.

It is important to facilitate effective drilling when positioning the cut and de-

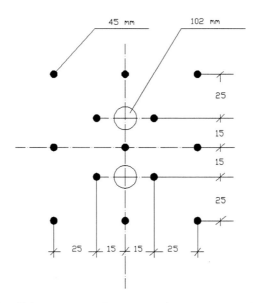

Figure 8: Parallel cut - Large diameter with two empty holes (102 mm)

signing the drilling pattern. An even distribution of the drilling work between the drills is necessary for making the drilling time as short as possible. The loading of the cut is most effectively done from the floor

$$\text{Number of lifter holes} = \frac{\text{Tunnel width} + 2 \times \text{offset of contour holes} + 1}{\text{Distance between holes}}$$

Suggested values for burden and distances between holes based on practical experience and rock of medium blastability:

Borehole diameter 34-38 mm:

Contour holes:	Burden 0.6-0.8 m	Distance between holes:	0.5-0.8 m
Next row:	Burden 0.9 m	Distance between holes:	0.8 m
Lifter Holes:	Burden 0.8 m	Distance between holes:	0.7 m

Borehole diameter 45 mm:

Contour holes:	Burden 0.8-1 m	Distance between holes:	0.7-1.0 m
Next row:	Burden 1.0 m	Distance between holes:	1.1 m
Lifter Holes:	Burden 1.0 m	Distance between holes:	1.0 m

4   Explosives – Loading

4.1   Definition

Any chemical compound, mixture or device the primary or common aim of
which is to cause a chemical reaction involving an extremely rapid expansion
of gases usually associated with heat.

4.2   General

Originally tunnels and rock caverns were almost exclusively blasted with explosives in cartridges. More and more the packaged explosives are going to be
substituted by bulk emulsions in tunnels, using borehole diameters of 45 mm
and larger.

4.3   Emulsion explosives

It has lately been demonstrated that emulsion slurry explosives have valuable
qualities in tunnel work both with respect to blasting and loading techniques
and from an environmental point of view.

The slurry explosives have been through a dramatic development during the
last 30 years. When they initially became available they were intended for
large boreholes in open pit mines. Today the emulsion slurry explosives have
taken over and enlarged their field of application to include extensive use in
smaller quarries, road cuttings and tunnels.

The slurry explosives of today can be most readily divided into the following
two types:

- Watergels

- Emulsions

It is common to both types that they are composed of raw materials such as
nitrates (ammonium, calcium and sodium), oils (fuel oils and other mineral
oils), water and special chemicals. The emulsion is a mixture of components
which are not by themselves defined as explosives. The mixture becomes an
explosive during the loading process (cartridges or bulk) when the density is
reduced by adding a gasing agent.

In simple terms the difference between the two types is that watergel has very
small droplets of fuel oil suspended in a nitrate solution whilst emulsion has

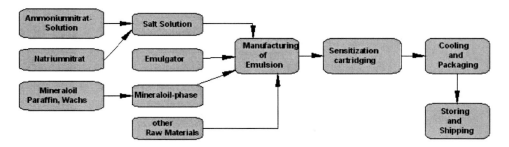

Figure 9: Production of emulsion explosives

microscopically fine droplets of nitrate solution suspended and surrounded by a continous oil phase.

The contact between the fuel oil and the nitrate is much better in an emulsion which gives a more complete chemical reaction with more energy released, higher speed of detonation and less emission of noxious gases which is an important factor when working underground.

Both watergels and emulsions have a high degree of water persistance. Another important factor is the higher level of safety in use, transport and during production compared with ore conventional explosives. Pumpable emulsion explosives have been tried in tunnels with good results in the last few years.

Advantages:

- Reduced emission of noxious gases

- Higher level of safety

- Increased water resistance

- The same type of explosives can be used in all holes

Disadvantages:

- The equipment is more complicated and requires trained operators

Emulsions are water resistant and pumpable explosives. This makes it possible to vary the quantity to be loaded from one hole to the next. The lifters and the production holes need a maximum of explosive energy whilst the contour holes require considerably reduced loading. This is achieved by mechanically operated withdrawal of the loading hose. The hose is pulled back at a certain rate while the pump is delivering an even amount of emulsion per unit of time. The parameters are adjusted to give the required amount of emulsion per meter of borehole. Safety is optimized by gasing the emulsion during loading. Storage and transport hazards of the explosive are eliminated.

For initiation of this pumpable emulsion a booster will be needed. There are two possibilities of boosters:

- 10 − 15 cm of 80 g detonating cord

- Austin Powder mini booster

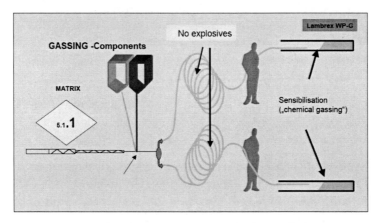

Figure 10: Austin Powder pump technology - Overview

Figure 11: Austin Powder pump technology - Loading process with two hoses

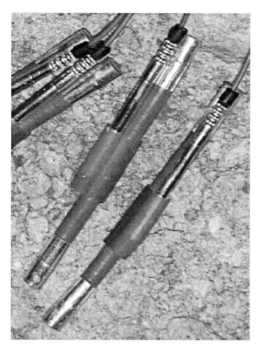

Figure 12: Booster of 80 g detonating cord inclusive detonator

Figure 13: Austin Powder Mini Booster

4.4 Nitroglycol-based explosives (dynamites)

Nitroglycol is produced from nitrieracid and glycol. The advantage from nitroglycol against nitroglycerine is that it is liquid up to $-22°C$. Nitroglycerine in dynamites were developed from Alfred Nobel in 1862.

Mixing together nitroglycol, ammoniumnitrate, nitrocellulose and other raw materials gives gelatinous exlosives also called dynamites. This explosives have the best performance but the nitrogen oxides (NO and NO_2) are very malicious.

With research and development on emulsion explosives in the early 1980's the use of dynamites was reduced in tunnel blasting and now they are only used as primer.

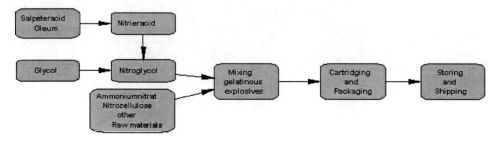

Figure 14: Production of dynamite explosives

4.5   Loading

The fan cut and wedge cut are almost universally loaded with dynamite and well tamped to achieve a high loading density. The production holes are loaded with a high energy explosive like dynamite for the bottom charge and with a weaker explosive in the rest of the hole.

The lifter holes are loaded with dynamite partly because of the generally wet conditions at floor level and partly to produce a muckpile which is easy to handle. The contour holes are loaded with emulsion explosives in small diameter cartridges of 22 mm and with a dynamite cartridge at the bottom of the holes as an initiator. This will reduce fracturing of the rock in the contour zone.

5   Initiation

5.1   General

90% of the detonators used underground in Austria are of the electric type, only 10% are non-electric. The non-electric detonators are primarily Shock-star and similar products. Electric detonators are available in different groups classified according to sensitivity. In Austria there are two different groups of detonators available. In group 'two' you have the highest protection against accidential firing by extraneous electricity. They are mostly used in tunnels when loading from electric/hydraulic rigs. This detonators have a critical current for circuit of 4.5 amp.

The firing pattern must be designed so that each borehole or group of boreholes are given as good break-out conditions as possible and a minimum of burden. It is also important that the rock to be blasted at each delay interval has sufficient room for break out.

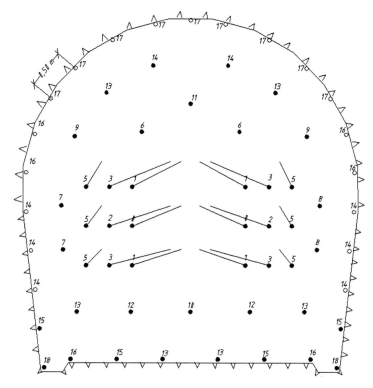

Figure 15: Full face firing pattern of a small cross section tunnel (18 sqm)

5.2 The cut

The delay intervals in the cut must be sufficient for the rock to be blown out of the hole before the next hole is fired. The delay time should be between 50 and 100 milliseconds. Soft rocks require longer delays than hard rock. The Austin millisecond detonators have a delay time of 25 or 80 milliseconds. This means that only every second or third delay number of the 25 ms detonators shall be used.

5.3 The production holes

The firing pattern of the production holes shall be arranged to give the best possible break for each separate hole

5.4 The contour

The contour holes along the wall are fired with the same delay number to achieve as much interaction as possible between charges. This is refered to as smooth blasting of the contour or trim blasting. The same applies to the contour holes in the roof which are fired with a higher delay number than the contour holes along the wall.

5.5 The lifters

The lifter holes are fired with later delay numbers than the holes in the roof. The last holes to be fired are the holes in the corner.

References

[1] Johansen, J.: Modern Trends in Tunnelling and Blast Design, Edition 2000

Injection for waterproofing and ground improvement in underground construction

Knut F. Garshol

Degussa Construction Chemicals, Zürich, Switzerland

Abstract: Grouting in rock poses high requirements on the grout to appropriately penetrate into the open joints and to be sufficiently persistent against flowing water. The mechanical properties of the several grout types such as cement and chemical grouts are presented and discussed with respect to their injection and to their effect on the rock mass.[1]

1   Introduction

1.1   Reasons for grouting in tunneling

Tunnel excavation involves a certain risk of encountering unexpected ground conditions. One of the risks is the chance of hitting large quantities of high pressure ground water. Also smaller levels of ground water ingress can cause problems in the tunnel or in the surroundings. Water is the most frequent reason for grouting in tunnels. Ground water ingress can be controlled or handled by drainage, pre-excavation grouting and post-excavation grouting.

Rock or soil conditions causing stability problems for the tunnel excavation is another reason for grouting. Poor and unstable ground can be improved by filling discontinuities with a grout material with sufficient strength and adhesion.

1.2   Short explanation of the Subject

Pressure grouting in rock is executed by drilling boreholes of suitable diameter, length and direction into the rock material, placing packers near the borehole opening (or some other means of providing a pressure tight connection to the borehole), connecting a grout conveying hose or pipe between a pump and the packer and pumping a prepared grout by overpressure into the cracks and joints of the rock surrounding the boreholes.

---

[1]This paper is part of the book "Pre-Excavation Grouting in Rock Tunnelling" by Knut F. Garshol and is here published with the kind permission of the author.

In tunnel grouting, there are two fundamentally different situations to be aware of:

- Pre-excavation grouting, or pre-grouting, where the boreholes are drilled from the tunnel excavation face into the virgin rock in front of the face and the grout is pumped in and allowed to set, before advancing the tunnel face through the injected and sealed rock volume. Sometimes such preexcavation grouting can be executed from the ground surface, primarily for shallow tunnels with access to the ground surface area above the tunnel.

- Post-grouting, where the drilling for grout holes and pumping in of the grout material take place somewhere along the already excavated part of the tunnel, because of unacceptable water ingress.

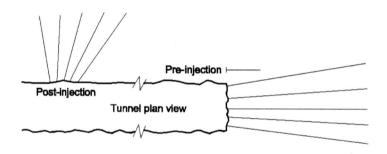

Figure 1: Pre-excavation grouting and post-grouting

The purpose of tunnel grouting in a majority of the cases is ground water in-flow control. Improvement of ground stability may sometimes be the main purpose, but will more often be a valued side effect of grouting for ground water control. Cement based grouts are clearly used more often than any other grout in tunnel injection, but there are also a number of useful chemical grouts available.

Pressure grouting (injection) into the rock mass surrounding a tunnel, is a technique that has existed for more than 50 years, and it has developed rapidly during the last 15 to 20 years. Much of the development into a high-efficiency economic procedure has taken place in Scandinavia. Pressure injection has been successfully carried out in a range of rock types, from weak sedimentary rocks to granitic gneisses and has been used against very high hydrostatic head (500 m water head), as well as in shallow urban tunnels.

The effect of carrying out grouting works ranges from close to drip free tunnels (around 1 l/min per 100 m of tunnel, [3] and [5]), to ground water ingress reduction dictated by practical and economical considerations (like specified acceptable remaining ingress in the order of 30 l/min per 100 m, [2]). It must be emphasised already at this stage that post-injection in this context is only a supplement to pre-injection. This important aspect of tunnel grouting will be explained later in this Chapter.

1.3 Traditional cement based grouting technology

Pressure grouting into rock was initially developed primarily for hydro power dam foundations and partly for general ground stabilisation purposes. For such works there is normally few practical constraints on the available working space. As a result grouting was mostly a separate task, and could be carried out without affecting or being affected by other site activities.

The traditional cement injection techniques were therefore applicable without too much of a disadvantage. The characteristic way of execution was:

- Extensive use of Water Pressure Testing (WPT) on short sections of boreholes (3-5 m), for the mapping of rock conditions and water conductivity (Figure 2). This process involves carrying out water pressure tests at regular intervals along the borehole to see what the overall water loss situation is i.e. which sections of the borehole are watertight and which sections allow the water to escape. The results were used for decision making regarding cement suspension mix design like water/cement ratio (w/c-ratio by weight), and to choose between using cement or chemical grouts.

- Use of variable and mostly very high w/c-ratio grouts (up to 4.0) and grout to refusal procedures, the latter expression meaning that grout is pumped into the rock until the maximum pre-determined pressure is reached and no more goes in.

- Use of Bentonite in the grout, to reduce separation (also called bleeding) and to lubricate delivery lines.

- Use of stage injection (in terms of depth from surface), low injection pressure and split spacing techniques (new holes drilled in the middle between previous holes). One way of stage injection involves drilling to a certain depth and then injecting the grout and next to re-drill and deepen the same hole and repeating the process. Split spacing as described above is also a form of stage injection.

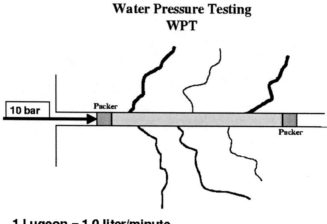

Figure 2: Water pressure testing of borehole

The typical effect of the above basic approach was that injection operations were quite time consuming - WPT every 5 m; pumping of a lot of water for a given quantity of cement; the need for counter pressure (i.e. pumping grout until the rock would take no more) causing unnecessary spread of grout; holding of constant end pressure over some time (say for 10 minutes) to compact the grout and squeeze out surplus water; slow strength development and complicated work procedures; it all added up to a long execution time. The low maximum pressure normally allowed to avoid any prospect of lifting the ground in which the grouting was being carried out (typically less than 5 bar, or with a relation to rock cover at the packer placement point), reduced the efficiency of the individual grouting stages leading to more drilling and injection steps, to reach the required sealing effect. See Figure 3.

In conclusion: The traditional cement injection technique, as described above and for the reasons given, is rather inefficient when considering the time necessary and the resources spent in reaching a specified sealing effect. This is especially the case when considering working from a tunnel face, where the rock cover and limited free surface area allow the use of fairly high pressure without the same risk of damage.

1.4   Rationale for the increase in the use of pressure grouting

In the last 20 years, pressure grouting ahead of the face in tunnels (referred to as pre-grouting, or pre-injection), has become an important technique in modern tunnelling works. There are a number of reasons for this:

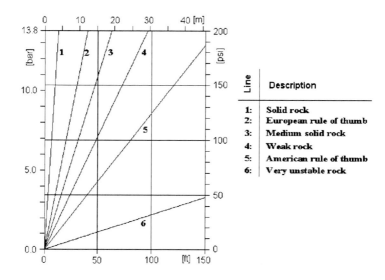

Figure 3: Relation between rock cover and admissible grouting pressure [8], (Typical for dams and other foundation grouting)

- Limits on permitted ground water drainage into tunnels are now frequently imposed by the local authorities, due to environmental protection and sometimes to avoid settlement above the tunnel. Settlement may cause damage on the surface, e.g. to infrastructure like buildings, roads, drainage pipes, supply lines and cable ducts. See example in Figure 4.

Figure 4: Northern Puttjern drained by Romeriksporten Tunnel in Oslo (Photo SCANPIX)

- The risk of major water inrushes, or of unexpectedly running into extremely poor ground, can be virtually eliminated (due to systematic probe drilling ahead of the face, being an integral part of the pre-grouting technology). It should be noted that if the excavation process hits a major water feature (because it was not detected and not pre-grouted), then water ingress has to be sealed in a post-grouting situation. This process is not only time consuming and expensive, but also far less effective than pre-grouting or pre-injection. In difficult situations, it can be close to impossible to succeed.

- Poor ground ahead of the face can be substantially improved and stabilised before exposing it by excavation. This improves the face area stable standup time, thus reducing the risk of uncontrolled collapse in areas of poor ground.

- Risk of pollution from tunnels transporting sewage, or other hazardous materials, can be avoided or limited. This is because once the ground has been treated by pre-injection it becomes less permeable so such hazardous materials cannot freely egress from the tunnel.

- Permanent sprayed concrete tunnel linings are increasingly being installed. The savings potential in construction cost and time is substantial, being the main reason for the increased interest in permanent lining shotcrete technology. Such linings cannot be produced with satisfactory quality under wet (running water) conditions, and ground water ingress control by pregrouting might become necessary.

With modern tunnelling drill jumbos even very hard rock can be penetrated at a rate of 2.5 to 3.0 m/min. In other words, the cost of probe drilling to guard against sudden catastrophic water inflows is now low. At the same time it should be noted that a large number of projects experience such catastrophic situations and are often stopped for months. Such events are extremely expensive and time consuming. It is then quite strange that the low insurance premium of limited probe drilling is not paid, to avoid the consequences of future possible huge water inrushes. This is especially so, when considering that if such conditions are identified ahead of the tunnel face, they can be treated successfully at a fraction of the cost and time spent if blasting into it. A list of examples could be made long, and some are shown in Table 1. (expanded by the author based on Fu *et al*, 2001).

Project name	Length (km)	Ingress (m³/min)	GW head (bar)	Location
Pinglin	12.8	10.8	20	Taiwan, R.O.C
Yung-Chuen	4.4	67.8	35	Taiwan, R.O.C
Central (East Portal)	8	18.6		Taiwan, R.O.C
Seikan	53.8	67.8		Japan
Semmering pilot	10	21		Austria
Gotthard Piora pilot	5.5	24	90	Switzerland
Isafjordur	9	150-180	6-12	Iceland
Abou	4.6	180	22	Japan
Luengchien trailrace	.8	81		Taiwan, R.O.C
NorthWest Himalaya	10	72		India
Access Oyestol		5	50	(flow in 1 borehole) Norway
Kjela (Bordalsvann)		15	23	Norway
Ulla Forre, Flottene		40		Norway

Table 1: Some examples of water inrushes at the tunnel face [6]

1.5   Some comments on Post-grouting

Grouting behind the tunnel face (post-grouting), should normally be used in tunnelling as a supplement to pre-grouting, to seal off remaining spot leakages if necessary. This will be especially necessary if the pre-grouting has not produced the required average tightness within a given section of the tunnel. It is interesting to observe that post-grouting becomes far more effective, when the same area has already been pre-injected. The normal problem of leakage points shifting from one tunnel location to another, without really sealing them off, is mostly avoided.

It has been repeatedly experienced in a number of projects that post-grouting alone seldom can produce the targeted result, or only after prohibitive use of resources. When a certain level of tightness is specified, it cannot be overemphasised that pre-injection has to be carried out. This is because this process seals the open joints in the rock before the water starts to flow, whereas with post-grouting the water has started to flow into the tunnel and the joints have to be blocked with the water flowing through them. One of the problems that has to be faced with post-grouting is grout wash-out . A study summing up some Norwegian projects indicates that the time and cost of reaching a specified result by post-grouting, may be 30 to 60 times higher than by pre-grouting [20]. A translation from Norwegian of the two last sentences of page 3 of this reference reads:

"However, it is recommended in cases where large water inrushes can be expected and especially at high ground water head, to carry out probe drilling ahead of the face and to carry out pre-grouting if large water flow is detected. Based on experience the cost of stopping water ingress by post-injection is 30-60 times higher than that of using pre-injection."

Other experienced engineers may be using different figures to illustrate the extra cost of using post-grouting exclusively, like 2-10 times more. An accurate figure does not exist so the important point to note is the general agreement that post-grouting is extremely expensive and complicated.

When pumping a grout into rock, the flow of the grout is governed by the principle of least resistance. The shortest flow path in post-grouting, offering least resistance, is very often leading back into the tunnel. To achieve spread of grout into the rock volume, backflow has to be stopped first. Furthermore, if a potential backflow path also carries flowing water, obviously the injected grout will suffer dilution and wash-out effects. The more water, the higher pressure and the larger the flow channels are, the more difficult it will be to seal them off. These are the very reasons for the dramatic cost difference presented in reference [20]. See also Figure 5.

1.6   New time saving methods and materials technology

The characteristic situation in all modern tunnelling is that the speed of tunnel advance is decisive for the overall economy. This fact is closely linked to the very high investment in tunnelling equipment, causing high equipment capital cost. Added to this is the fact that the limited working space at the tunnel face normally allows only one work operation to take place at a time.

The face advance rate is decided by the number of hours available for actual excavation works (other factors kept constant). Time spent for pre-injection will normally have to be deducted from this available excavation time. One hour of face time typically has a value of more than US $ 1000 and it is evident that the efficient conduct of all activities at the tunnel face is a priority. From this, it can be seen that injection in a tunnelling environment is fundamentally different from injection for dam foundations and ground treatment from the surface. This is the main reason why the technical development in tunnel injection has been different to other types of rock injection.

Because of the need to save time (and therefore cost), technical specifications for routine tunnel grouting cannot be loaded with tests and investigative techniques. If it is required to carry out extensive water pressure testing in stages in all holes, if core drilling is made part of the routine drilling from the face, if joint orientation and crack openings have to be checked by camera

Figure 5: Very difficult to seal by post-injection (Photo: Peter Town)

etc., and all this is linked to a complicated system of decision-making during execution of grouting, the sum may be termed overkill. Such research related activities can not be made part of the routine grouting works if cost and efficiency has any priority. The additional down-side is that such over-zealous procedures will probably not improve the end result at all.

The last 15 years has led to the development of a number of new cement based products for injection. Typically, these cements are ground much finer and may offer more suited setting and hardening characteristics. In most cases, these cements are combined with admixtures or additives to provide entirely new cement grout properties and substantially improved penetration into cracks. When combined with working procedures that are adapted to the new materials properties the efficiency increase is substantial. Even though these new cement products are more expensive than standard Portland cements, they are still very competitive, compared to most traditional chemical grouts (refer to Figure 6).

Cement based grouts remain the material of first choice for pressure grouting in tunnelling. This is due to the low volume cost, availability, well documented

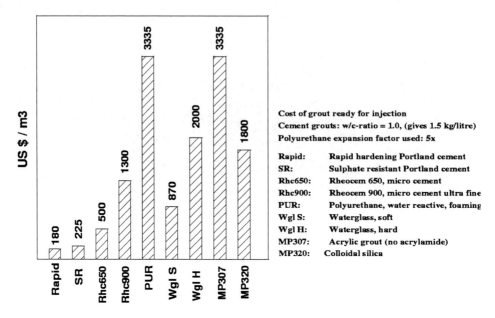

Figure 6: Relative material volume cost of various injection products

properties and experience and environmental acceptability. However, the wide range of available chemical grouts offers a useful supplement to cement grouts, especially when the tightness requirements are strict. Chemical grouts can penetrate and seal cracks that cementitious grouts will not enter.

2 Grouting into rock

2.1 Particular features of rock (compared to soil)

Rock materials and soils are fundamentally different in terms of the behaviour of water flow and the effect of injecting any kind of grout into the ground.

Soils possess a wide variation in particle sizes, layering, compaction, porosity, permeability and a number of other parameters. However, at basic level soils consist of particles and permeability is directly linked to the pores (spaces or voids) between the particles.

Between discontinuities, most rock materials, on the other hand, are practically impermeable for water and grouts. Leakage and conductivity is therefore linked exclusively to discontinuities within the rock mass. It is necessary to understand and accept this important difference between soil and rock, to be able to correctly evaluate all aspects of pressure grouting in rock tunnelling and to understand why the approach has to be different to soil injection techniques.

When comparing rock and soil, the similarities and differences are primarily governed by how scale is being treated. It is important to understand and take account of the effects of scale to reach correct solutions and answers. If the conditions within a whole mountain are considered, the average permeability of the rock mass can be measured and evaluated by the same methods as are normally used for soils (a similarity). The reason for this is that the overall rock mass fragmentation creates very small block sizes (similar to particles in the soil case) compared to the whole mountain volume under consideration and the whole mass can be treated with a reasonable approximation as being homogeneous.

In comparison, when considering the rock volume for the first few meters around a tunnel and along a few meters of its length, single joints and channels will govern and dominate the pattern of water conductivity and grout take. In such a randomly chosen limited rock volume, the joints and channels can show water conductivity many orders of magnitude different to the mountain average permeability (a difference). To use the term permeability in the same sense as for soils, therefore can be highly misleading. In a perfectly homogeneous sand volume of a given permeability one could, as an example, calculate 300 l/min water ingress into a 100 m tunnel length. If mentally assuming that the sand is impermeable but with an inserted steel pipe through the sand into the same tunnel, the pipe used to feed 300 l/min of water (which could be an illustration of the hard rock water conducting channel situation), the average permeability would be the same. However, the two situations are certainly totally different in practical terms if looking for water sealing solutions.

The permeability term is being used to estimate and illustrate ground water flow conditions on an overview level also in hard rock (large scale average), which is an acceptable approximation for this situation. However, on a detailed level in hard rock, the term permeability is not applicable and practical decisions made based on an assumed permeability will mostly turn out to be totally wrong.

For injection in soils the following indications have been given by Karol [11]:

$k = 10^{-6}$ or less   not groutable
$k = 10^{-5}$ to 10^{-6} groutable with difficulty by grouts under 5 cP viscosity and not groutable for higher viscosities
$k = 10^{-3}$ to 10^{-5} groutable by low-viscosity grouts but with difficulty when viscosity is more than 10 cP
$k = 10^{-1}$ to 10^{-3} groutable with all commonly used chemical grouts
$k = 10^{-1}$ or more   groutable by suspended solids grout

Based on the previously mentioned differences between soil and rock, the above guidelines will not be applicable in most rock materials. With WPT results in boreholes as basis for calculation of permeability in rock even section lengths as short as one meter could easily indicate permeability between one and three orders of magnitude too low. In addition, the fact that rock injection in tunnelling allows the use of much higher injection pressure (often 10 times more) will change the practical limits of what is groutable and not.

In a rockmass it is evident that the characteristics of jointing will be of major importance for any grouting program. The variation of joint properties and water conductivity in different types of rock is actually extreme and a discussion of this subject is outside the scope of this paper.

However, some examples can be given to illustrate the importance of the subject and to draw the attention to some effects of typical conditions found in rock. Perhaps the most extreme water conductivity situation that can be found is in limestone, where carst features occur. These are solution channels in limestone formations that can create huge caverns and literally allow a subterranean river. Even when the channel has a typical diameter of only one meter, the water flow conditions into a tunnel intersecting it would be catastrophic.

Hard rock materials like gneisses, granites and quartzites, will often show unweathered jointing patterns at depth, that may result in a substantial total leakage potential. Such jointing can be quite easy to inject and seal. Local fault areas, especially major shear zones in the same kind of bedrock, may contain a lot of fine material and clay gouge. Such zones will often show no leakage at all due to the fines, but if there are local water bearing channels, they may be difficult to find and complicated to seal off. Uncontrolled running water in such channels may lead to flushing out of fine materials from the zone, resulting in increasing flow over time. Such effects also depend on the ground water pressure.

Weaker beds like shales, limestones, mudstones, sandstones and some metamorphic rocks are often jointed and layered to a considerable degree. A high number of water bearing small cracks may in total produce substantial leakage. A complication for a successful injection program in such rock conditions,

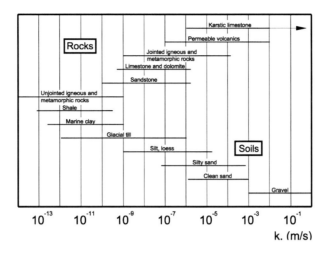

Figure 7: Average permeability of soil and rock

is often the wide variety of joint filling materials that can be found. Such joint fillings tend to inhibit grout penetration and distribution and the fill materials are sometimes squeezed around by the grout being injected. See Figure 7.

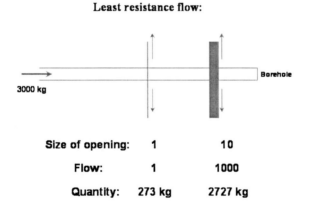

Figure 8: Effect of conductivity contrast on grout flow into open joints

In most rockmasses, the main problem for pressure grouting is the non-uniform conditions caused by localised geological features. In a borehole of some meters length there will, in most cases, be a mixture of joints, cracks and channels, between water-tight sections. Any fluid pumped into such a borehole will inevitably follow the path of least resistance. The effect of this is that a given volume of grouting material, may follow a very conductive opening at fairly low pressure, to a distance much greater than expected and beyond what is effective. At the same time, there will be very limited penetration into other openings (due to low pressure and material lost into the main channel). This problem can and very often does lead to unsatisfactory grouting results, and/or increased cost, due to increased number of grouting stages and too high material consumption to achieve the required result. See Figure 8.

In a rock type with only one clearly dominating joint set, where one would expect water leakage and grout penetration to generally flow along these joint planes, this is only partly going to occur. Observation of the nature of water ingress in TBM excavated tunnels (where additional blasting cracks are not obscuring the natural conditions), clearly demonstrates that channels within joint planes are the typical situation. This is well demonstrated by leakages appearing as concentrated point jets from somewhere along the joint intersection with the tunnel periphery.

Experience from post-grouting in tunnels further supports the idea of channel leakage and channel conductivity as the normal mechanism of water transmission in jointed hard rock. When a water flow clearly is originating from an identified joint plane, that can be observed crossing the tunnel periphery, drilling can be performed to cut through the joint plane at a suitable depth and angle, with the purpose of getting direct contact to the water flow. Often, a number of holes need to be drilled across the joint plane, to actually hit the water leakage. The reason is obvious - most of the joint plane is dry and the water flows through a limited channel within the plane. When drilling for water flow contact, it is of course much more difficult to hit a pipe than a plane.

An example can be given from the Norwegian hydro power project Kjela (1977). At tunneling length 1800 m from access Tyrvelid, direction Bordalsvann, the tunnel hit a water inrush of 15 000 l/min at 23 bar pressure. As could be clearly seen in the tunnel, more than 90% of this inrush came from one concentrated channel located within a shear zone.

2.2   Handling of rock conductivity contrast

For reasons of time-related cost and adequate cover tunnel pre-injection requires relatively long boreholes (10 to 30 m) and injection through one packer placement near the opening. In such length of borehole there will normally be conductivity contrast along the hole, sometimes this contrast may be extreme. With a large conductivity contrast and grout flow in direction of least resistance it is necessary to take steps to reduce the negative effects of this normal situation. See Figure 9.

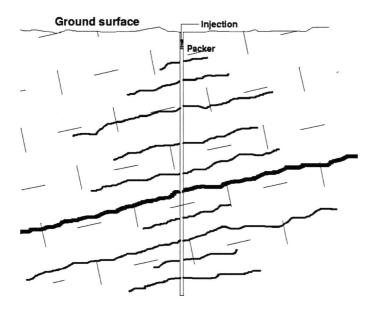

Figure 9: Large conductivity contrast

The problem is that chemical grouts will flow into on the large openings at low pressure doing nothing to seal smaller openings. Cement grouts will have the same tendency and grout to refusal gives excess material consumption. Stable cement grout and suitable procedures can counteract the problem to a large extent and increase efficiency. The best way of illustrating how to deal with conductivity contrast is by using an example situation (Figure 9), treated with traditional grout to refusal technique and alternatively with stable cement grout and dual stop criteria.

2.2.1 Description of typical Grout to refusal procedure

Start of grouting with a w/c-ratio 3.0, high grout flow at very low pressure and assuming that 90% of the flow goes in the largest channel. Standard procedure is to reduce the w/c-ratio in steps when the pressure is not increasing. One may assume that after 3.5 hours spent injecting say 4000 kg of cement and reaching the maximum allowed pressure (for the specific conditions), the following situation has been reached:

- cement has travelled in the largest channel to a maximum distance of 350 m from the borehole (which is far beyond the useful spread).

- the grout pressure increased gradually, especially during the last part of the injection time.

- grout permeation into medium and small cracks is only in mm-scale. This is caused by a long period under low pressure and clogging of the cracks by filter cake development. Also, when the pressure finally increases the grout used has a low w/c-ratio and higher viscosity and so would not permeate that easily.

- some of the injected grout has separated, leaving residual openings and conductivity.

2.2.2   Stable grout of micro cement using dual stop criteria

The whole injection can be executed with a fixed 1.0 w/c-ratio and a low viscosity of 32 s Marsh cone flow time, using a thixotropic grout. Also in this case 90% will flow into the largest channel at very low pressure. After one hour of injection time the stop criterion of 1500 kg per hole has been reached (pressure still low). (In most practical cases like this the micro cement procedure would utilize a limited volume at w/c-ratio 1.0 like 250 litres, before changing to 0.8 and later to 0.6. This will not change the described examples other than in making the micro cement alternative work even better). The established situation may be assumed to be as follows:

- micro cement has travelled on the largest channel to a maximum distance of 125 m from the borehole (which is also beyond the useful spread). This shorter distance is primarily caused by less cement being pumped. See also next bullet point.

- some penetration has been achieved into medium and small cracks due to the grout stability, low grout viscosity and smaller particle size.

If assuming that the hole length used was 12 m, the next step would be to drill a new neighbour hole with the same length. This would take about 5 to 10 minutes with modern drilling equipment. Injection can now be done into the same area (large channel blocked by first stage injection) and penetration will therefore be into medium and small cracks at a higher injection pressure. It can be assumed that it takes 30 minutes to inject 500 kg cement when the allowed maximum pressure has been reached.

2.2.3 Comparison of the two procedures

	Traditional OPC grouting	Stable micro-cement grouting
Time spent	3.5 hours	1 hour 40 minutes
Materials spent	4000 kg OPC	2000 kg micro cement
Injected	1 stage, basically one crack	2 stages, large and small cracks
Result	ineffective	mainly effective

The micro cement alternative using half the material and less than half the execution time, has achieved the following result improvements compared to the OPC procedure:

- Since two grouting stages have been executed, the achieved rock tightness in the first few meters around the holes is much better. Other reasons for better tightness are the fact that the grout viscosity was always very low, the grout was stable (so no re-creation of channels due to bleeding) and the maximum cement particle size would typically be 1/4th of the OPC.

- The grout durability and strength is substantially better because of the lower w/c-ratio and no use of Bentonite in the mix.

It would be an option to also execute two stages using OPC and then the result could of course be improved. However, this would then again take additional grout and additional time and experience shows that the result would be poorer. The cost of extra cement and even more important, extra time will normally cause substantially higher overall cost for a poorer result, using an OPC and grout to refusal technique.

2.3 "Design" of grouting in rock tunnels

Design of grouting in rock tunnels means essentially the development and specification of drilling patterns, the grout materials to be used and the methods and procedures to be applied during execution. These are the variables which can be controlled by engineers, geologists or specialists and which are varied according to local actual conditions in the tunnel, with the purpose of achieving a specified result. The outcome cannot be accurately predicted because of the nature of the technique and the lack of details about ground conditions. Nobody can directly observe what happens in the ground during

Figure 10: Overall situation in km-scale, GW and rock parameters variation in cm-scale

injection, other than the indirect signs and effects on water ingress and by inspection after excavating through the grouted rock volume.

Even the evaluation of carefully controlled full scale tests can be difficult. The uncertainty about unforeseen changes in ground conditions from one test location to the next cannot be accurately quantified. However, most of the principles of pre-grouting have been developed through and are supported by the results of several thousand tons of grout injection in tunnelling and the understanding of the principles is not so much guesswork as it is sometimes claimed to be.

The word "design" probably needs to be commented upon to clarify what it means in the context of tunnel grouting. The need for such a clarification arises from the difference to the normal understanding of the term when used in structural design.

Design of a bridge or a high-rise building will include the necessary drawings, materials specifications and structural calculations to define the dimensions, the geometry, the load bearing capacity, the foundations and the general layout of the object to be built. The whole analysis has to be based on the given physical surroundings, the owner s requirements regarding service loads, service life expectancy and other features or limitations that are applicable.

In the case of a tunnel grouting operation many will expect the above principles to be applicable so far as the design process is concerned. However, the reality is that it is not possible to design the work with precision in advance of it being carried out so it is nothing like the design process referred to in the previous paragraph. The design of tunnel grouting operations is based upon the best estimates of the average permeability of the rock through which the tunnel is to be driven. The design will usually include calculations of the likely water ingress, drawings showing matters such as the depth, angle and pattern of the intended drilling, execution procedures covering all aspects of the operation and the materials specification, so as to aim at satisfying the required water tightness of the tunnel. There is no question of drawings being produced showing what the finished job will look like or to give accurate dimensions for the result.

The pre-investigations for rock tunnel projects can never give sufficient details about the rock material and the hydrogeological situation for the full length of the tunnel, so as to allow a bridge design approach. Furthermore, the calculation methods available are not refined enough to accurately analyse the link between the required result and the necessary steps to produce it. To further compound the problem it must also be admitted that even if assuming that a mathematical model would be available there is no chance that all the materials parameters could be measured, accurately quantified and input to such a model. See Figure 10.

The basic design for the grouting operation as referred to above has to be applied in practice on an empirical, iterative observational design-feedback basis (monitoring of results) as described below:

- once the "watertightness" requirements are defined, the project data and all available information about rock conditions and hydrogeology can be analysed and compared with those requirements. This often includes indicative calculations of potential ground water ingress under different typical situations. Based on empirical data (previous pre-injection tunnel project experience) a complete pre-grouting method statement can then be compiled. However, irrespective of how elaborate this method statement (or "design") is and whatever tools and calculations are employed to produce it, it will not be more than a prognosis. This prognosis will express how to execute the pregrouting (under the expected range of ground conditions), what sequence of steps to take to meet the required tightness of the excavated tunnel.

- during excavation the resulting tightness in terms of water ingress achieved can be measured quite accurately. This means that it is possible to move to a quantitative comparison between targeted water ingress and the actual result and accurately pinpoint if the situation is satisfactory or not. If the results are satisfactory, the work will continue without changes, and only a continued verification by ingress measurements will be necessary.

- If the measured water ingress rate is too high, this information will be used to decide on how to modify the design to ensure satisfactory results compared to the requirements for the remaining tunnel excavation. This may have to be executed in stages, until satisfactory results are achieved. Excavated tunnel sections which do not meet the requirements of the specification will have to be locally post-grouted until the overall result for such sections are acceptable unless it is possible to compromise on the water tightness requirements.

2.4   Fluid transport in rock

The permeability of a material expresses how readily a liquid or a gas can be transported through the material. Darcy s Law is based on laminar flow, an incompressible liquid with a given viscosity and is valid for a homogenous material [17]:

$$v = ki$$

where: v ... flow velocity
k ... coefficient of permeability
i ... hydraulic gradient

The requirement of a homogenous material is never satisfied for jointed rock materials, and then only when the volume being considered is big enough. Normally, the term joint permeability, or even better conductivity should be used.

The coefficient of permeability can be measured in the laboratory, using the above given formula of Darcy:

$$q = kAi$$

where: K ... absolute permeability (m^2)
k ... coefficient of permeability (m/s)
μ ... dynamic viscosity (mPa·s) or cP
ν ... kinematic viscosity (m^2/s)
g ... 9.81 m/s^2
γ ... volume weight of the liquid (N/m^3)

For testing of rock mass conductivity through bore holes, the unit Lugeon is the most frequently used. Lugeon (L) is defined as the volume of water in litres that can be injected per minute and meter of borehole at a net over-pressure of 10 bar (see Figure 2).

The Lugeon value needs interpretation and cannot be considered in isolation. If measurement has taken place over a bore hole length of say 10 m, then there is in principle, always the chance that all the water has escaped through a single leakage location. This means, that if the same borehole had been measured in 0.5 m increments, nineteen of these would have had a L-value of zero, while one would be 20 times the above measured average.

To avoid possible extreme differences between Lugeon values resulting from a single measurement over a long bore hole (10 to 30 m) and the real value over shorter segments (like 1 m), technical specifications sometimes requires that the Lugeon value calculation length is set to 5 m for all borehole measuring lengths longer than 5 m.

The following table illustrates the different units discussed above:

Materials/Units	Lugeon	k (m/s)	K (m^2)
Fine sand	100	10^{-5}	10^{-12}
Jointed granite	0.1	10^{-8}	10^{-15}

2.5 Practical basis for injection works in tunnelling

Pre-injection in tunnelling may have various purposes and may be carried out under quite variable geological- and hydrogeological conditions. All these factors will strongly influence how to execute pre-injection in a given case. However, there are a few basic, practical facts of common nature when at a tunnel face that must be part of any pre-injection planning and execution.

At a tunnel face, typically there is limited working space and the logistics may be an added problem. Mostly, working operations at the face are sequential and very little can be executed in parallel. To keep the cycle time short and the rate of tunnel face advance high, it is extremely important that all work sequences are as rapid as possible, with as small disturbance and variation as possible and with a smooth change from one operation to the next. Of course, this is decisive for the cost of the tunnel, since the time related expenses are running whether there is face advance, or not.

One very important aspect of tunnel face injection activity must be emphasised. In general, injection into jointed rock materials is not an easily preplanned activity. Pre-investigations may have yielded a lot of general information, but very little on a detailed level. On the other hand, a lot of specific and detailed information is generated during drilling of holes and during execution of the injection itself. The temptation on the part of planners and designers to create very elaborate working procedures, lots of tests, voluminous record keeping and tight supervision is therefore very strong. If such a tendency is not checked this can generate very complicated and time consuming decision procedures. Lots of detailed information must be processed with clear lines of authority and decisions must be made regarding the influence on further future work operations. It is very easy to end up in a situation where the good technical intentions in the end are detrimental to the purpose of the exercise.

Elaborate WPT procedures with the purpose of choosing the type of grout are frequently relied upon far beyond the technical merit of the procedure. Plotting of experience data to check on the possible correlation between grout take and originally measured Lugeon value will be very disappointing. One example of such data is shown in Figure 11. All such efforts that the author has come across are similar to what is shown in this figure.

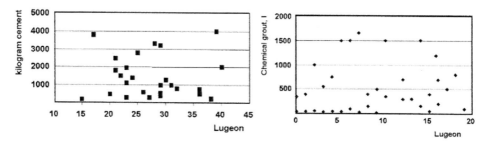

Figure 11: Correlation between measured L-value and grout consumption [1]

Another basic aspect of pre-injection must be kept as a part of planning and operation. Regardless of the reason for the pre-injection, as long as it has to do with water leakage control, 100% water cut off is not realistic or cost effective. Also, when the requirement is expressed as some pre-defined rest leakage level in the excavated tunnel, it is not possible to accurately hit the targeted leakage rate (see Design of grouting in rock tunnels). Whether applying very strict requirements (like 2 l/min. and 100 m tunnel), or ten or twenty times this, experience shows a wide result variation around the target value. There is no known, feasible way of substantially improving this lack of accuracy. There are therefore clear limitations to what volume of refinements and sophistication that are reasonable and productive to undertake in the injection procedures.

This may seem negative and may be understood as a complete lack of control of the injection process. It is not, because of two main factors:

- Water ingress measurements in already excavated tunnel parts will tell where the criteria are not met, how much off the results are, under what conditions and resulting from which resource allocation already used in probe drilling and pre-injection. The same goes for the tunnel sections with satisfactory results. This information and its evaluation can be continuously fed back to the at-face execution for necessary correction of procedures. Experience shows that the target results will then be more closely reached and with a more optimal use of resources.

- In those areas where the criteria are not met, post injection can be undertaken, normally starting with the highest-yield leakage points. This technique is very efficient when pre-injection has already been carried out (otherwise, leakage would normally just be moved around). Because of the actions described in the above bullet point, the need for post injection is quickly reduced and the final result will meet specified requirements.

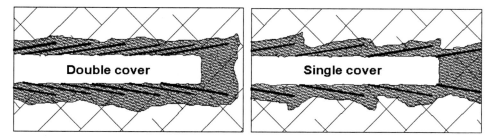

Figure 12: Double and single cover grouting

Since it is generally so much more efficient to execute pre-injection it is also better to start out a little on the conservative side with the works procedures and later to relax the approach if appropriate as experience is gained. When requirements are tight and the potential consequences of not meeting criteria are serious, it will often be best to simply decide on pre-injection as a routine systematic activity using a double cover approach (see Figure 12). The rationale is that if probe drilling in most cases will lead to pre-injection, then this separate activity and decision-making can be saved thus simplifying the procedures and increasing the efficiency.

In less strict situations e.g. with maximum allowed final water ingress of 30 l/min in 100 m tunnel and no consequences in the surroundings of the tunnel, a limited overlap of typically 5 m per 20 m probing length (25%) can be used. In this type of case normally probe drilling will be used to provide basis for decision about where to actually execute pre-injection. Sections of the tunnel that are relatively dry without injection will then be passed through using probing alone. Where injection has to be done this can be the so-called single cover approach, see Figure 12.

2.6   Grout quantity prognosis

Practically all pre-grouting in hard rock tunnelling is based on the use of cement (OPC or micro-cement). In special cases, like in ground conditions with clay and other fine materials on the jointing planes and/or when the required tightness cannot be reached with cement only, chemical grouts may become necessary as a supplement. There is no experience basis available for the use of predominantly chemical grout, to illustrate typical consumption. However, in the case of cement injection this can be done.

Also in the case of cement only grouting, the required quantity will depend on a large number of factors and any estimate made in advance will be inaccurate. The main influence factor is the rock conditions (properties of the jointing), where a limited number of large open channels will tend to require more cement than cm-scale joint spacing producing frequent drips (distributed rain in the tunnel). Other important factors are required tightness, static head of ground water, tunnel cross section and even the type of cement and injection methodology applied.

From sub-sea tunnelling with systematic probe drilling and partly with systematic pre-grouting there are average consumption values from quite variable Scandinavian conditions between less than 20 kg/m tunnel to more than 250 kg/m. As an extreme case the Bjoroy sub-sea road tunnel stands out with a section of about 500 m tunnel length consuming 2000 kg/m. Target water ingress level was typically 30 l/min per 100 m tunnel.

When evaluating empirical data covering such a wide range it can be useful to view the data on a probability basis. Three different figures can be used to illustrate the experience data available from Norwegian sub-sea road tunneling:

1. Minimum average consumption, with 5% probability that the average will be lower than this figure

2. The probable average consumption

3. Maximum average consumption, with 5% probability that the average will be higher than this figure

The minimum can be expressed as 15 kg/m tunnel, probable value is 50 kg/m and the maximum average is 500 kg/m. These values are roughly representative of predominantly hard rock types (but not only granitic rock materials) and the tunnel length would have to be 1000 m or more to yield a reasonable average. Obviously, such figures can only be taken as an illustration of what has been experienced before and they can not be transferred directly and accurately to new projects in other ground conditions.

It must be mentioned for clarity that the figures are averages including tunnel sections that needed no grouting at all.

3   Functional requirements

3.1   Requirements and ground water control during construction phase

Based on the above evaluations on the functional requirements for the tunnel, the tunnel design and execution and its relation to the surroundings; a number of issues have to be decided regarding the ground water control program. The difficult problem to solve is how to satisfy the requirements during all stages of construction and operation of the tunnel.

One requirement that is frequently overlooked, is the water ingress rate during the construction phase of a project. If the tunnel will be constructed in an urban area and ground water lowering could cause settlement damage to infrastructure on surface, then it is not enough to plan for a final water tight permanent stage lining. It may take weeks and months between the time of exposing the ground at the face, until the water tight lining has been established in the same location. Meanwhile, substantial volumes of water may have entered the tunnel, lowering the ground water level. Frequently it is too late to prevent settlement and damage, if the ground water level comes up to normal again some months later. The situation illustrated in Figure 13 is such a case.

The only available tunnelling technique that can keep the ground water in-leakage near zero, is the Earth Pressure Balance Machine (EPBM), full face mechanical excavation using a pressurised shield and gasketed concrete segment installation. Such machines are for soil excavation and are limited to shallow depths (typically less than 15 m).

In hard rock tunnelling this alternative is not available, even if a TBM and concrete segments are used for the excavation and support. Without pre-injection the leakage volume could locally become far too large, between the time of exposure and the time of segment erection and efficient annular space backfilling. With a serious local water inrush at hand, such segment handling and grouting would also be very difficult.

Ordinary in-situ concrete lining, even with waterstop in the construction joints, has hardly any influence on the water ingress level, as shown by ingress

Tunnel drainage effect

Very small water ingress to the tunnel =>
several meter pressure drop in the sand =>
pore pressure loss and settlement in the clay

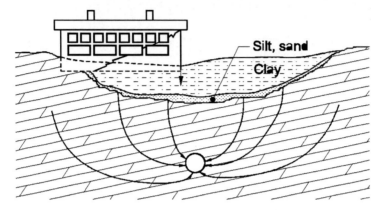

Figure 13: Particularly sensitive situation. Examples of several dm settlement

levels of 10 to 40 l/min per 100 m tunnel. [10] In the Oslo area this is typically the ingress rate for an unlined and not pre-injected tunnel. Concrete lining with careful high pressure grouting of the interface to the rock was still quite successful. Concrete lining with PVC membrane gave acceptable result, but was also not completely water tight. [10] Two important conclusions can be drawn:

1. A concrete lining will frequently be in place too late to prevent permanent damage on surface

2. Concrete lining with contact grouting or PVC membrane will typically cost more than an extensive pre-grouting operation, achieving about the same final result

Therefore, there are situations where probe drilling and pre-grouting has to be executed to meet the requirements of ground water control during the construction phase.

3.2   Measurement of water ingress to the tunnel

As described under "Design" of grouting in rock tunnels, there is no way of directly and accurately linking the grouting works effort and the final water ingress result. The result has to be monitored, corrected if necessary by doing postinjection as needed and by correcting the way the pre-grouting is being executed.

To be able to accurately determine what is the water ingress result after injection, this has to be measured for pre-defined tunnel lengths. Depending on the requirements and the necessary accuracy of these measurements, tunnel lengths could be 10 m, 100 m or even more. The normal way of measurement is by dams in the tunnel floor (especially prepared and sealed to avoid wrong results) equipped with an overflow V-notch (or any other defined shape that can be used to calculate the flow rate).

One alternative is the 90° V-notch where the height of water above the bottom of the notch can be used in the formula:

$$q = 43 \cdot 10^{-6} \cdot h^{2.5}$$

where q is flow of water in l/s, h is the water height in mm above the bottom of the V-notch. For quick reference, the diagram in Figure 14 can be used.

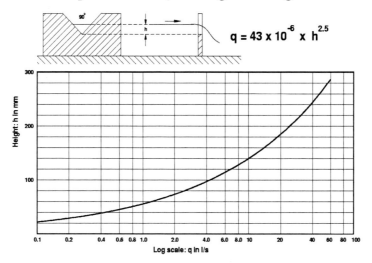

Figure 14: Measuring water flow rate by V-notch overflow

4 Cement based grouts

4.1 Basic properties of cement grouts

4.1.1 Cement particle size, fineness

Any type of cement may be used for injection purposes, but coarse cements with relatively large particle size, can only be used to fill bigger openings. Two important parameters governing the permeation capability of cement, are the particle size and particle size distribution. The average particle size can be expressed as the specific surface of all cement particles in a given quantity. The finer the grinding, the higher is the specific surface, or Blaine value (m^2/kg). For a given Blaine value, the particle size distribution may vary and the important factor is the maximum particle size, or as often expressed the d_{95}. The d_{95} gives the sieve dimension where 95% of the cement particles will pass through (and conversely, the remaining 5% of the particle population is larger than this dimension. The maximum particle size should be small, to avoid premature blockage of fine openings, caused by jamming of the coarsest particles and filter creation in narrow spots.

The typical cement types available from most manufacturers, without asking for special cement qualities are shown in Table 2.

Cement type / Specific surface	Blaine (m^2/kg)
Low heat cement for massive structures	250
Standard Portland cement (CEM 42.5)	300-350
Rapid hardening Portland cement (CEM 52.5)	400-450
Extra fine rapid hardening cement (limited availability)	550

Table 2: Fineness of normal cement types (largest particle size 40 to 150 μm)

The cements with the highest Blaine value will normally be the most expensive, due to more fine grinding.

Table 3 gives an example of particle size distribution of cements commonly used for pressure injection. Please note that the actual figures are only indications, since the table is based on single measurements, from single cement samples.

From an injection viewpoint, these cements will have the following basic properties:

- A highly ground cement with small particle size, will bind more water than a coarse cement. The risk of bleeding (water separation) in a suspension created from a fine cement is therefore lower and a filled opening will remain more completely filled.

- The finer cements have a quicker hydration and a higher final strength. This is normally an advantage, but causes also the disadvantage of shorter open time in the equipment. High temperatures will increase the potential problems of clogging of lines and valves. The intensive mixing required for fine cements, must be closely controlled, to avoid heat development caused by the friction in the high shear mixer, and hence even quicker setting.

Percent passing				
μm	Norwegian rapid hardening (RP38)	Spinor A12 Ciments d'Origny	W650 Blue Circle	Swedish Injection cement Degerhamn
1	7.0	12.9	10.1	12.6
3	22.0	59.0	31.2	30.4
5	32.0	82.5	45.2	40.5
10	50.7	98.3	68.8	55.4
15	65.6	100.0	86.5	66.6
20	76.9		95.5	73.8
25	86.3		99.1	80.9
32	95.6		100.0	90.4
40	99.6			97.1
50	100.0			99.8
μm				
90%<	27,3	6.4	16.6	31.6
10%<	1.3	0.5	1.0	0.9
Average particle size	9.8	2.5	5.9	7.9

Table 3: Particle size of some frequently used injection cements

The finer cements will give better penetration into fine cracks and openings. This advantage will only be realized as long as the mixing process is efficient enough to separate the individual particles and properly wet them. In a pure cement and water suspension, there is a tendency of particle flocculation after mixing, especially with finer cements, and this is counter-productive. It is commonly said that the finest crack injectable, is about 3× the maximum particle size (including the size of flocculates). For standard cements, this means openings down to about 0.30 mm, while the finest micro cements may enter openings of 0.06 mm.

The question is sometimes raised, what is the definition of microcement. Unfortunately, this question cannot be answered based on any kind of internationally accepted agreed definition and it is left to common practice and case by case identification. As an informative indication of a minimum requirement to apply the term micro cement, the following suggestion can be used:

Cement with a Blaine value > 600 m^2/kg and minimum 99% having particle size < 40 μm.

The above "definition" fits quite well with the International Society for Rock Mechanics reference [4.1]:

> "Superfine cement is made of the same materials as ordinary cement. It is characterised by a greater fineness ($d_{95} < 16$ μm) and an even, steep particle size distribution."

An example of a micro cement just satisfying the superfine definition can be found in [7, Chapter 11: MBT Injection Materials]. The Rheocem$^{©}$ 650 has a Blaine value of 650 m^2/kg and the particle distribution shows 94% < 15 mm.

The effect of water reducing admixture (or dispersing admixture) when mixing a micro cement suspension can be seen in Figure 15 [4]. It is quite evident that the reduction of d_{85} by the use of a dispersing admixture from about 9 μm to 5 μm will strongly influence the penetration of the suspension into the ground. If these figures are put into the soil injection criteria of Mitchell [16], a good injection result with this cement without admixture could be achieved in a soil with $d_{15} > 0.22$ mm. With admixture the same result could be obtained in a soil with $d_{15} > 0.12$ mm. Also in rock injection the effect would be significant.

Another important effect of water reducing admixtures is the lowered viscosity at a fixed w/c-ratio. The effect of lower water content is improved final strength of the grout, but more important is lower permeability and better chemical stability. The compressive strength of a pure water and cement mix using a standard OPC is about 90 MPa at w/c-ratio of 0.3 (which will be far too stiff to be used for normal injection). Already at a w/c-ratio of 0.6 the strength will drop to 35 MPa and when using grout mixes with w/c-ratio above 1.0 the strength is finally in the range of 1.0 MPa and less. (Rheocem$^{©}$ 650 microcement with 1.5% admixture reaches 10 MPa compressive strength after 28 days). More important in cases with even higher water content is that the permeability is pretty high and the strength is so low that if any water flow takes place, it can lead to mechanical erosion and chemical leaching out of hydroxides (hydration products from cement reacting with water).

4.1.2   Bentonite

Bentonite has traditionally been used on a routine basis in combination with cement for grouting of soil and rock. The reason to do so was the strong tendency of standard cement to separate when suspended in water, enhanced

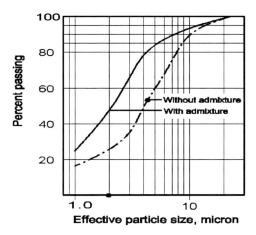

Figure 15: Dispersing effect of an admixture when using micro-cement [4]

by the normal use of water cement ratio > 1.0. Bentonite can be used to reduce the bleeding in such grouts and a standard dosage of 3 to 5% of the cement weight has a strong stabilising effect.

Bentonite is a natural clay from volcanic ashes and the main mineral is montmorillonite. There are two main types:

- Sodium-bentonite (Na-)

- Calcium-bentonite (Ca-)

Mostly the sodium-Bentonite is used as an additive in cement grouts, because it swells to between 10 and 25 times the original dry volume when mixed in water. The particles resemble the shape of playing cards and will adsorb water on the particle surfaces, thus stabilising the grout mix. The particles also sink very slowly within the suspension because of the shape. See Figure 16.

With the traditional cement grouting methods and materials Bentonite had its place. However, in combination with micro-cements it is normally not necessary and will mostly be of disadvantage. One reason is that a typical d_{95} particle size of Bentonite clay is around 60 μm. This is two to three times larger than what is found in good micro-cements and will reduce the penetration achievable by a given cement. The shape of the particles are also a negative property in this respect. Modern micro-cement grouts can be made with very low viscosity and limited or no bleeding if combined with chemical admixtures and the Bentonite use is therefore unnecessary and negative for the result.

The final strength of the grout is not important in most cases. However, at high ground water head, or when a ground stabilisation effect is valuable,

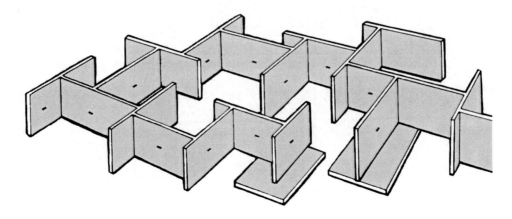

Figure 16: Idealised structure of Bentonite clay after dispersion in water.

the use of Bentonite at normal dosage will reduce the grout strength by 50% and more. This is avoided when using modern admixtures in micro-cement grouts, without sacrificing stability or penetration.

4.1.3   Rheological behaviour of cement grouts

Cement mixed in water as an unstable suspension, or as a stable paste (in terms of water separation), behaves according to Bingham's Law. Water and true liquids have flow behaviour according to Newton's Law. These laws are as follows (see Figure 17):

Bingham's law: $\tau = c + \eta \ dv/dx$
Newton's law: $\tau = \eta \ dv/dx$
where: τ ... flow shear resistance (Pa)
                 η ... viscosity (Pa·s)
                 dv/dx ... shear velocity (s^{-1})
                 c ... cohesion (Pa)

When a stable grout has a very low w/c-ratio, or when ground mineral powder or fine sand has been added, the grout may also have an internal friction. To cover this property, Lombardi has proposed the following rheological formula [13]:

$$\tau = c + \eta \ dv/dx + p \tan \varphi$$

where: p ... internal pressure within the grout
         φ ... angle of internal friction of the grout

A true liquid will flow as soon as there is a force creating a shear stress. Water in a pipe will start flowing, as soon as there is an inclination. A liquid with a higher viscosity than water will also flow, but at a lower velocity.

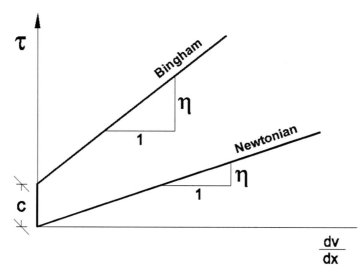

Figure 17: Rheological behaviour of Newton and Bingham fluids

A cement suspension or paste, will demonstrate some cohesion. The difference to liquids is that the cohesion has to be overcome, for any flow to be initiated. If the internal friction is negligible, the paste will thereafter behave in a similar manner as a liquid. The rheological parameters of cement suspensions can be influenced by w/c-ratio, by chemical admixtures, by bentonite clay and by other mineral fillers. As an example, it is possible and often useful, to create a grout with a high degree of thixotropy. This means a paste with a low total flow resistance while being stirred or pumped, but shortly after being left undisturbed, it shows a very high cohesion.

4.1.4 Pressure stability of cement grouts

For the purpose of controlling grout flow in the ground (to be able to place the grout where it is wanted), the control of the rheological parameters of the grout is vital. In this context, there is one more factor that is very important: The grout stability under pressure, which is not tested or reflected by the normal check on bleeding. The best way to illustrate the point, is to consider two different grouts, both having w/c-ratio low enough for zero bleeding. If these grouts are filled into a container with a 45 μm micro filter in the bottom and subjected to pressure, two things may happen:

- The grout with a good stability, will loose a very small quantity of water through the filter and the thickness of dried out and compacted grout on top of the filter will be very thin. The main part of the grout under pressure remains uninfluenced.

- A grout with a poor stability will over the same time loose much more water through the filter and a thick layer of dried out and compacted grout will be found on top of the filter. If the pressure is high enough and the grout stability is very poor, all the grout volume may be dried out and compacted.

The standard method for testing the pressure filtration coefficient (Kpf) is the American Petroleum Institute (API) Recommended Practice 13. The coefficient is defined as the volume of water lost using the API filter press divided by the initial volume, divided by the square root of the filtration time in minutes, using a 6.9 bar pressure (100 psi).

When squeezing out a small quantity of water from the grout at the injection front (which is well simulated by the API pressure filtration test), internal friction will quickly increase the flow resistance enough to stop further permeation. This will cause the pump pressure to increase, having the effect of more water being pressed out and a rapid development of a plug. This will happen more readily with a poor stability grout and often in positions where the openings are much bigger than the rule of thumb 3× maximum particle size.

Practical project experience and results support the above views and it is likely that the grout stability is much more important for the permeation of a cement grout, than some limited difference in particle size.

4.1.5   Use of high injection pressure

High injection pressure has proven very successful in achieving low water ingress levels, far better than what was within reach some years ago. As described above the pressure filtration is an important factor in this and it is clear that the best effect will be reached with a combination of a high grouting pressure (above 50 bar) and a grout with a low filtration coefficient.

Furthermore, the grouting pressure, when high enough, will also dilate the cracks and joints of the rock formation and thus increase penetration by increasing the opening size. If high pressure is used without careful consideration of the consequences, it will be possible to cause damage. Especially, be careful not to use very high pressure in combination with large grout quantities in a single continuous pumping sequence.

Keil *et al* [12] used stable microfine cement injected into a granite formation. This full scale injection test was well instrumented and revealed opening and closure of fracture zones by as much as 100 micron. Analysis of specimens from the grouted formation revealed penetration into cracks as fine as 20 micron. It should be noted that the grout used had a relatively high viscosity of 44 seconds Marsh cone time.

4.1.6 Grout setting characteristics

Ordinary Portland cements will typically show the following ranges of initial and final setting times and 24 hours uniaxial compressive strength (ISO mortar, MPa):

Initial set: 140 to 240 minutes
Final set: 190 to 240 minutes (10 to 20 MPa at 24 h)

A typical high early strength (rapid hardening) Portland cement in comparison:

Initial set: 80 to 180 minutes
Final set: 150 to 240 minutes (15 to 30 MPa at 24 h)

From a practical point of view, initial setting time cannot be made much shorter, without potential problems of build up in equipment and clogging of material lines. Of course, it is possible to use admixtures to control the open time, which is covered separately.

One has to be aware that final set has limited relevance compared to strength, or hardening created by cement hydration. Under field conditions in tunnel injection, the ground water hydrostatic pressure may be in the range of 10 to 50 bar (sometimes even higher). If a high hydrostatic pressure is combined with fairly large openings, then sufficient time has to be allowed at the end of an injection stage, before any drilling or blasting into the same area. Otherwise, a puncture may occur and injected cement and water is flushed back into the tunnel, thus destroying the work carried out and creating a hazardous condition. The necessary time will also depend upon the w/c-ratio of the injected grout and at w/c-ratios substantially greater than 1.0, whether or not compaction has been carried out by standing end pressure.

In extreme cases, the necessary waiting time may be as long as 24 hours (above 30 bar and openings larger than 50 mm). In a more moderate case (pressure of 5 to 25 bar and maximum openings up to 25 mm), a waiting time in the range of 10 to 15 hours should be sufficient. As a rule of thumb, keep in mind that the compressive strength reached at a w/c-ratio of 1.0 will be only 25 to 30% of that at w/c-ratio of 0.4 and a further reduction to only 5% at w/c-ratio of 2.0. From this, it should be quite obvious that it will pay off to carefully evaluate the situation under difficult conditions and control the w/c-ratio.

4.2   Durability of cement injection in rock

There is a large volume of hard rock tunnelling with extensive use of pre-injection as part of the tunnel design and as the sole measure of permanent ground water control. The primary experience basis is probably in Scandinavia, where Norway alone has close to 100 km of sub-sea tunnelling. Even though some of these tunnels go down to as much as 260 m below sea level and the grout injection works carried out are of a permanent nature, there is no report indicating that grout has degraded.

The Norwegian Public Roads Administration operates 17 sub-sea tunnels of various different ages (the oldest tunnel goes to Vardø island and was commissioned in 1981) excavated through quite variable ground conditions. In fact, the general trend reported is a slow reduction of water ingress over the years as opposed to any sort of degradation in the grout which, of course, if it occurred would lead to an increase as opposed to a reduction in the water ingress. Melby [15] presents a paper dealing with 17 different projects totalling 58.6 km of tunnelling. A comparison of water ingress at the time of opening and measurements made in 1996 shows the average ingress in 1996 to be only 62.9% of the ingress recorded when the tunnels opened. None of the tunnels showed an increase in the leakage rate.

The Norwegian national oil company Statoil constructed three pipeline sub-sea tunnels amounting to a total of 12 km, going down to 180 m below sea level. The tunnels are crossing Karmsundet, Førdesfjord and Førlandsfjord. During more than 15 years of operation Statoil has recorded the energy-consumption expended in pumping of ingress water from the deepest point in the tunnels to sea level discharge. Statoil states that there has been no increase in ground water ingress, since the energy consumed has not increased [18].

Very important for the quality and durability of cement grouts is the w/c-ratio and whether the grout is stable or segregating. Modern grouting technology in tunnelling means stable grouts and thus also w/c-ratio below a certain limit, depending on the type of cement and the admixture used. This view is supported by the ISRM, Commission on Rock Grouting, Final Report, which states in [9, Chapter 4.2.6]:

Stable or almost stable suspensions contain far less excess water than unstable ones. Hence, grouts with a low water content offer the following advantages:

- During grouting:
    - higher density, hence better removal of joint water and less mixing at the grouting front

  – almost complete filling of joints, including branches

  – the reach and the volume of grout can be closely delineated

  – grouting time is shortened because little excess water has to be expelled

  – the risk is reduced that expelled water will damage the partially set grout

- After hardening:

  – greater strength

  – lower permeability

  – better adhesion to joint walls

  – better durability

5 Chemical grouts

Chemical grouts consist of only liquid components, which leads to a quite different behaviour than cementitious grouts. Chemical grouts behave like Newtonian fluids, demonstrating viscosity but no cohesion (see Figure 17). Therefore the penetration distance from a borehole and the placement time for a given volume, only depend on the viscosity of the liquid grout and the injection pressure used. Chemical grouts available include silicates, phenolic resins, lignosulphonates, acrylamide and acrylates, sodium carbomethylcellulose, amino resins, epoxy, polyurethane and some other exotic materials.

For practical purposes there are two main groups of chemical grouts available:

- Reactive plastic resins

- Water-rich gels

The reactive resins may be monomers or polymers that are mixed to create a reaction (polymerisation) to a stable three-dimensional polymeric end product. When short reaction times are used, normally such products are injected as two-component materials, mixing taking place at the injection packer upon entry into the ground. At longer reaction times even two component materials can be injected by a one component pump. Mixed batches only have to be small enough and with long enough open time to be injected before the polymerisation reaction takes place. Such products will not be dissolved in water, but they may react with water. For proper reaction and quality of the end product the right proportioning of the components is important. Two component pumps must function properly for this to be satisfied.

The gel forming products are dissolved in water in low concentrations and the liquid components therefore show a very low viscosity (often almost as fluid as water). When the polymerisation takes place, an open three dimensional molecular grid is created, which binds a lot of water in the gel. The water is not chemically linked to the polymeric grid, but is locked within the grid by adsorption.

5.1   Polyurethane grouts

5.1.1   General

Polyurethane grouts (PU-grouts) are reactive plastic polymers having a wide range of properties for practical applications. Polymers are giant molecules that are produced by joining smaller molecules (monomers) in a so called step growth (or condensation) polymerisation into e.g. polyurethane products. Products with repeating units of $NHCO_2$ are called polyurethane. A simple example reaction is shown below:

$$CH_3\text{-}N\text{=}C\text{=}O \;+\; HO\text{-}CH_2\text{-}CH_3 \;\;\rightarrow\;\; CH_3\text{-}NH\text{-}CO\text{-}O\text{-}CH_2\text{-}CH_3$$
$$\text{Isocyanate} \;+\; \text{Alcohol} \;\;\rightarrow\;\; \text{Urethane}$$

The reaction products may be rigid or soft, pore free or foamed up to 30 times the volume of the liquid components and the reaction time may vary between seconds and hours. The viscosity of mixed product, before reaction has started and also the speed of reaction, are both quite sensitive to temperature. There are products for practical application to be injected as a single component, as well as twocomponent systems. The properties of a product are mainly governed by the choice of different basic raw materials. Most systems can be modified by the use of added catalysts and other chemicals that influence the behaviour of the product.

The very wide range of possible PU-grout properties offers an advantage to the specialist, to tailor make a material for specific purposes. For the normal end user, this complexity can be quite frustrating, because it will be difficult to sort out which commercial product is the best one for an intended application.

The normal way to offer some flexibility, without complicating matters too much, is that manufacturers will offer a limited number of standard products, with a set of properties for a range of typical situations. On the basis of such a palette of standard products, it will be possible for solving special problems, to add modifications, by involving specialists on the job site.

The polyurethanes are formed by reaction of two components:

- Polyisocyanate (Diphenylmethane-diisocyanate, or abbreviated "MDI"). (There are also other isocyanates available, but these are more hazardous and should not be used for injection in any underground project).

- Polyalcohols (abbreviated "polyol")

The structure of a polyurethane molecule created from polyglycol is illustrated in Figure 18.

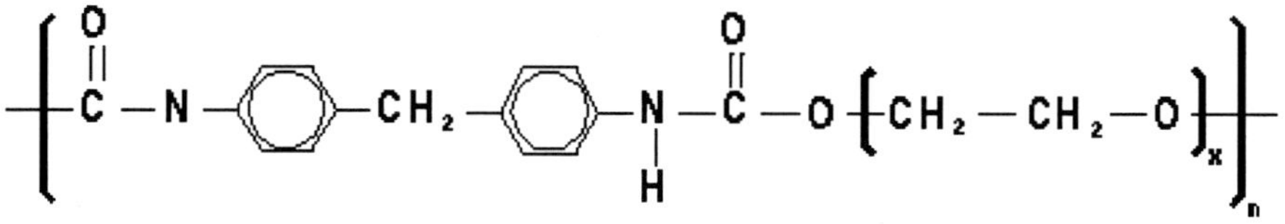

Figure 18: Polyurethane molecule

One very interesting part of the reaction is the effect of water. If some water is added to the polyol component, or the mixed components are meeting water after injection, a part of the isocyanate will react to polyurea and carbon dioxide (CO_2). This reaction takes place in parallel to the formation of polyurethane and the gas generates trapped bubbles, causing the formation of a closed cell foam.

In most cases under ground, there is a need of combined effects from an injection, like water flow cut-off and ground consolidation. The cost of materials is also important. The best consolidation is reached when there is very little foam reaction, but this will reduce the penetration into finer openings and the volume cost becomes high. At the other extreme, a very quick foaming to several times the original volume, will produce a low strength grout, that may be very effective for an initial cut-off of running water, but with little consolidation effect. A very porous grout will also not seal completely and subsequent water pressure build up, may compress the foam and increase the leakage again. The volume cost drops with the foam factor. The foam formation has the effect of self-injection of the PU-grout, because the CO_2-pressure developed can be up to 50 bar (temperature dependent). The penetration of the grout is therefore not only governed by pump pressure and by the product viscosity, but is also very much influenced by the foaming pressure.

The properties of the foam created will depend on the local conditions. When free foaming produces a volume increase of 30 times, in the ground the restricted volume increase will create pressure and less expansion. A typical average volume increase in rock injection at low pressure, is more like 5 to 10 times.

Polyurethane products have typically a high viscosity, which is a limiting factor for permeation into the ground. At room temperature a typical product viscosity is 200 cP, but it is possible to get as low as 100 cP. If the products are diluted by the addition of solvents it is possible to come down to about 20 cP, but solvents can cause health problems and environmental problems under ground.

5.1.2   Pumping equipment

For 2-component PU it is necessary to use a custom design 2-component PU pump. These are normally prepared for 1:1 ratio of the A and B components (by volume) and the whole set-up all the way up to the packer is shown diagrammatically in Figure 19.

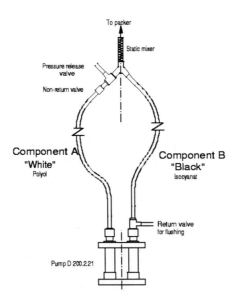

Figure 19: Pump and other accessories for 2-component PU

5.2 Silicate grouts

Sodium silicates have been used for decades as soil injection grouts. There are also examples of silicate injection in rock formations. The main advantage of silicate grouts is the low cost and the low viscosity. It may also be added that apart from the pH of typically 10.5 to 11.5 (causing it to be quite aggressive), there are small problems with working safety and health. Silicates are used for soil stabilisation or for ground water control.

Liquid silicate (also called waterglass) is produced by dissolving vitreous silicate in water at high temperature (900°C) and high pressure. The liquid is later diluted by water to reach a viscosity level that can be used for injection purposes in soil and fine cracks in rock. A normal injection grout will have a viscosity of about 5 cP and the gel produced is water rich, weak and somewhat unstable. Some syneresis will take place after gel has been formed in the ground (release of water from the gel and some shrinkage). Because of the low gel strength it will have limited resistance to ground water pressure, especially in cracks and joints that are relatively large. This can be seen in rock injection locally, where channels may be some cm wide, by a slow extrusion of gel over time.

The liquid silicate needs a hardener to create a gel. Acids and acidic salts will cause such gel-formation (like sodium bicarbonate, sodium aluminate), but today normally proprietary chemical systems will be used, showing much better practical properties with improved quality of the final grout. These products are mostly methyl and ethyl di-esters.

If the grouting is done as a ground water control of a permanent nature (several years), then silicates cannot be used. The syneresis is one of the problems in such an application, that can lead to new leakage channels over time, but also the chemical stability is questionable in many cases. For temporary ground water control for some months it will mostly be acceptable. In rock injection it will often be necessary to do cement-injection as a first step, to fill up the larger channels. The low pH cement-environment is very unfavorable for the durability of a silicate grout.

5.3 MBT colloidal silica

This product has no resemblance to the silicate systems described above. The colloidal silica is a unique new system with entirely new properties and can even be considered more environment friendly than cement (see also [7, Chapter 11]).

5.4   Acrylic grouts

The acrylic grouts came in use already 50 years ago and for cost reasons these
were based on acrylamide. The toxic properties of such products have over
the years stopped them from being used. The last known major application
was in the Swedish Hallandsasen tunnel, where run-off to ground water caused
pollution and poisoning of livestock. However, it is not necessary to include
this dangerous component in an acrylic grout.

Polyacrylates are gels formed in a polymerisation reaction after mixing acrylic
monomers with an accelerator in aqueous solution. In the construction industry, acrylic grouts are used for soil stabilisation and water proofing of rock.
Polymerised polyacrylates are not dangerous for human health and the environment. In contrast to that, the primary substances (monomers) of certain
products can be of ecological relevance before their complete polymerisation.
Injection materials polymerise very quickly - as a rule within some minutes.
Before the monomers completely polymerise, a considerable amount can be
diluted by the ground water, subsequently leading to contamination.

Because of such effects in practical injection works underground and because
of the working safety of personnel, the use of products containing acrylamide
(which is a nerve poison, is carcinogenic and with cumulative effect in the
human body) must be avoided.

Products are available that are based on methacrylic acid esters, using accelerator of alconal amines and catalyst of ammonium persulphate. These
products are in the same class as cement regarding working safety and can be
used under ground, provided normal precautions are taken.

Acrylic gel materials are very useful for injection into soil and rock with predominantly fine cracks. Normally such a product is injected with less than
20% monomer concentration in water and the product viscosity is therefore
as low as 4 to 5 cP. This viscosity is kept unchanged until just before polymerisation, which then happens very fast. This is a very favorable behavior
under most conditions. The gel-time can typically be chosen between seconds
and up to an hour.

The strength of the gel will primarily depend on the concentration of monomer
dissolved in water, but also which catalyst system and catalyst dosage that
is being used. The gel will normally be elastic like a weak rubber with a
strength of about 10 kPa at low deformation. An injected sand can reach a
compressive strength of 10 MPa.

If a gel sample is left in the open over time under normal room conditions,
it will loose the adsorbed water trapped within the polymer grid, shrink and

Figure 20: MEYCO MP 301 injected in sand

become hard. If placed in water, it will swell again and regain its original properties. In underground conditions this property of an acrylic gel will seldom represent any problem, but be aware that if an unlimited number of drying/wetting cycles must be assumed, then the gel will eventually disintegrate. The chemical stability and durability of acrylic gels are otherwise very good.

5.5 Epoxy resins

Epoxy products can have some interesting technical properties in special cases, but the cost of epoxy and the difficult handling and applicationare the reasons for very limited use in rock injection under ground.

Epoxy resin and hardener must be mixed in exactly the right proportions for a complete polymerisation to take place. Any deviation will reduce the quality of the product. The reaction is strongly exothermic and if openings are filled that are too large (width > some cm) the epoxy material will start boiling and again the quality will be reduced. Also for epoxy the viscosity is high, unless special solvents are used.

Working safety and environmental risk are additional aspects of epoxy injection that makes the product group of marginal interest for rock injection underground.

5.6   Combined systems of silicate and acrylic materials

In practical grouting it is quite normal to combine different grouts during the execution of the works. This will normally consist in reaching a certain level of tightness by the use of cement and then to finalise by some chemical grout. However, there are also products available where different chemical systems are combined into one commercial product.

Best known is combination of silicate and acrylic grout. The silicate component will lower the volume cost of the final product and the acrylic component will improve the chemical stability, reduce the syneresis and give a much stronger and more stable gel.

The product will be handled as a two-component material, where the hardener for the silicate is mixed into the acrylic monomer and the hardener for the acrylic grout is mixed with the silicate. When the two components are mixed, there will first be a silicate gel reaction, which is then followed by an acrylic gel formation to reinforce and stabilise the final gel.

The practical handling of such a system is a bit complicated and the use of such products is therefore quite limited.

5.7   Bitumen (asphalt)

In tunnel excavation it has happened a few times that extreme water ingress is exposed locally at the face. Such inrush can be catastrophic and will in most cases be extremely difficult to get under control or to seal off.

It has also happened that hydropower dams expose water channels from inside the water reservoir to downstream of the dam, causing severe water loss. Leakage like that can be extremely difficult to seal off, because it is mostly not an option to empty the water reservoir. The water pressure is therefore always present and the flow rate in the channels that need to be sealed, can be very high.

Typically, grouting of ordinary cement grouts in such situations is useless. The grout has no chance to set and is diluted and flushed out by the turbulent flowing water. Up to a certain limit, quick foaming polyurethane can be used for water flow cut-off, but there are situations where it will not work, especially at low temperature (slow reaction). From case reports it is known that a number of very innovative methods have been tried, like cement or concrete mixed with wood cuttings, bark cuttings, cellulose materials etc., and with all kinds of accelerators. Frequently failing to do the job.

As a last resort, heated liquid bitumen (asphalt) can be an alternative. The principle will be to use a selected quality of bitumen (roofing grade asphalt), that heated to a sufficiently high temperature (typically 200 to 230°C) has low viscosity allowing easy pumping. The softening point should be around 95 to 100°C. The output must be adapted to the water flow rate, the water head and the distance from the injection point until the downstream outlet point. However, the asphalt output may be less than 1% of the water leakage rate and still be effective. This is totally different to all sorts of cementitious grouting, where the grout flow rate must be able to displace the water to avoid washing out.

The ideal bitumen quality will rapidly change from an easily pumped fluid material to sticky, highly viscous and non-fluid asphalt at the water temperature. When injected into the water stream, the bitumen will rapidly loose its high temperature and rapidly and dramatically change its rheological properties. The bitumen gets sticky, will easily stop in narrow points in the water channel and can thus block the flow.

After a blockage has been achieved, it is always advisable to place some suitable cementitious grout to ensure a permanent and stable barrier.

At the Stewartwill Dam in Eastern Ontario, Canada, two concentrated leakage zones through the dam foundation were grouted by asphalt (combined with cement) [14]. The work was carried out with a full reservoir (about 6 bar water head). The first zone, grouted in 1983, yielded 13 600 l/min water leakage and this was reduced by more than 90%. The other zone, grouted in 1984, was 9 000 l/min and was reduced to virtually nil. It is interesting to note that both cases where executed in one day of grouting. Materials consumption was 6000 l asphalt and 5.7 m^3 sand (1983) and 3370 l asphalt and 2.8 m^3 sand (1984). An unsuccessful attempt in 1982, using cement and sand took 2 months and consumed 5600 bags of cement plus 73 m^3 sand.

The specialist contractor FEC Inc. carried out injection with asphalt in Pleasant Gap [19], near State College PA, USA. The grouting was running over 5 shifts and Figure 21 shows one of the water-flow exits after the 4th and the 5th shift.

Figure 21: Flow after 4th shift (left) and 5th shift grouting (right) (Photo P. Cochrane)

References

[1] Berge, K. O., Water control reasonable sharing of risk, Norwegian Tunnelling Society, Publication No. 12, Oslo 2002.

[2] Blindheim, O. T., Oevstedal, E., Design principles and construction methods for water control in subsea road tunnels in rock, Norwegian Tunnelling Society, Publication No. 12, Oslo 2002.

[3] Davik, K, I., Andersson, H., Urban road tunnels a subsurface solution to a surface problem, Norwegian Tunnelling Society, Publication No. 12, Oslo 2002.

[4] De Paoli, B., Bosco, B., Granata, R., Bruce, D. A., Fundamental observations on cement based grouts: Microfine cements and Cemill process, International Conference Soil and Rock Improvement in Underground Works, Milan, 1991.

[5] Erikson, A, Palmqvist, K., Experiences from the grouting of the Lundby tunnel , Proceedings of the Rock Mechanics day 1977, Swedish Rock Engineering Research Foundation, Stockholm, 1977.

[6] Fu, R., Sun, L. J., Wang, C. L., Catastrophic water inflow in the new Yung-Chuen Tunnel, Proceedings of the AITES-ITA 2001 World Tunnel Congress, Milan Italy, 2001, Vol. III, pp 143-150.

[7] Garshol, K.F. (2003): Pre-Excavation Grouting in Rock Tunnelling, MBT International Underground Construction Group.

[8] Houlsby, A. C., Construction and design of cement grouting, a guide to grouting in rock foundations, John Wiley and Sons, New York, 1990.

[9] ISRM (1995): Final Report of the Commission on Rock Grouting. International Society for Rock Mechanics.

[10] Karlsrud, K., Control of water leakage when tunnelling under urban areas in the Oslo region, Norwegian Tunnelling Society, Publication No. 12, Oslo 2002.

[11] Karol, R. H., Chemical Grouting, Marcel Decker, Inc., New York, 1983.

[12] Keil et al, Some new initiatives in cement grouts and grouting, 42nd Canadian Geotechnical Conference, Winnipeg, 1989.

[13] Lombardi, G., Deere, D., Grouting design and control using the GIN principle. Water Power and Dam Construction, Volume 45, No 6.

[14] Lukajic, B., Smith, G. and Deans, J., Use of asphalt in the treatment of dam foundation leakageStewartwill Dam , Seminar on issues in dam grouting, ASCE Spring Convention, Denver, Colorado, April 1985.

[15] Melby, K., Daily life of subsea rock tunnels construction, operation and maintenance , Proceedings of Workshop Strait Crossings Subsea Tunnels, Oslo, 1999.

[16] Mitchell, J.K., Soil improvement State of the Art Report , Proceedings X ICSMFE, Stockholm, vol. 4, pp 509-565, 1981.

[17] Norwegian Tunnelling Society, Handbook no. 1, Injection in rock, practical guidelines for injection strategy and methodology (in Norwegian).

[18] pers. comm. Operations Engineer Hans Ove Fostenes, Statoil, 1999.

[19] pers.comm. Paddy Cochrane, FEC Inc., Mascot, TN 37806, USA

[20] Stenstad, O., Execution of injection works (in Norwegian), Proceedings of Post Graduate Training Course sponsored by the Norwegian Chartered Engineer Association and the Norwegian Rock mechanics Group, Fagernes, Norway, 1998.

Exploration by core drilling

Martin Happel

Comdrill Bohrausrüstungen GmbH, Im Kressgraben 29, D-74257 Untereisesheim, Germany

Abstract: The various existing core drilling techniques for geotechnical site exploration are described and commented with respect to the quality of the extracted cores.

1 Introduction

There are many ways to get an idea of the nature of a rock behind it's face – the best way is to hold the rock itself in your hands. It's true information without any doubt – when you know how to get it. Thus, you should know a few things about geology and the tools how to treat this geology.

The most common mineral in earth's crust is quartz (SiO_2): it is hard and abrasive. Quartz is the enemy of the tool, of the bit, the corebit. Abrasivness causes wear. The only material to withstand this wear is diamond.

Core drilling is the story of attacking the rock – with the most expensive tools to conquer it – not to destroy it.

Figure 1: Core recovery; unscrewing the bit from a single tube core barrel; material: concrete

2   The coring system

Coring technology includes the following notions:

Bit: Short, threaded steel cylinder, on the drill front equipped with diamonds, fixed in a matrix.

Core barrel: Cylindrical tube to cover the core.

Drill rod: Enables the driller to bring bit and core barrel to any depth required. Thrust, revolution, torque, flush are supplied from the machine
to the top of the barrel.

Drilling machine: Power and control station for the drilling process.

Driller: Most important part in the drilling process. He will have to make
the best core quality. And he will give you the best explanation for the
core quality in the box.

3   Core barrels

3.1   Single tube core barrel

It is the simplest coring equipment. It consists of only one tube.

Figure 2: Single tube core barrel: Head, barrel, reamer, core lifter, bit

Bit: It has a thin cutting "face", kerf (8-10 mm) and is mainly applied for
very hard rock. Rotation and flush attack the core:

- Loose parts (possibly the most interesting ones) will be washed
away.
- The bit does not only drill the annular space, but also the washed-
away detritus from the core. This implies a reduction of its lifetime.

Main cutting diameters are 36 / 46 / 56 / 66 / 76 / 86 / 101 / 116 / 131 / 146 mm.

Reamer: Stabilizes the bit to keep the calliper. Integrated is a conical seat for the core lifter.

Core lifter: Splitted ring with conical, smooth outer shape and cylindrical, rugged inner shape. It acts as a trap when pulling the core barrel: The core lifter grips the core to extract it.

Core barrel: Receives the drilled core with lengths of 1.5-3.0 m. The core is covered but not protected.

Head: Is the link to the drillrod. As there is a huge variety in threads of drill rods (metric / DCDMA / GOST ...) it is important to know the right specification.

Core barrel types:

> **type B:** thin walled (see Fig. 3)
>
> **type Z:** thick walled (rather rare)

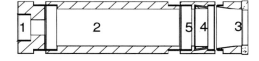

1. Core barrel head
2. Core barrel
3. Bit
4. Core lifter
5. Reamer

Figure 3: Drilling with single tube core barrels: only coherent and competent materials remain as mainly loose parts

3.2 Double tube core barrels

Tube-in-tube system

The inner tube protects the core. Rotation and mud flow do no longer attack the core. Even soft sections in the core can be recovered.

Outer tube system: It consists of bit, reamer, outer tube and core barrel head.

Inner tube system: It consists of core lifter case with core lifter, extension tube and inner tube with thread to core barrel head.

Figure 4: Cores of various qualities

Ball bearings in the core barrel head (swivel-type) allow that only the outer tube rotates. A fine machined push-pull system in the inner tube allows thin kerf bits; a thread connection is not needed.

Lengths of core barrels are mainly 1.5-3 m (single tube / double tube). Every completed core can only be recovered by (re)pulling the whole core barrel. Thus, the deeper you get, the longer and harder you have to work!

Core barrel types: T / T-2 / T-6 / T-6 S / K-3 / D

Sizes: 36 / 46 / 56 / 66 / 76 / 86 / 101 / 116 / 131 / 146 mm

The core quality can be improved by modifying the shape of the bit, using flushed core bits.

An additional tube (PVC) can be put in the inner tube to protect very sensible cores or to allow coring in contaminated areas, e.g. modified D 131×102 mm.

Splitted inner tubes also allow a good core recovery, e.g. T-6 S 131×102 mm.

Mud-press-core-barrel heads use their special design to pump out cores from the inner tube by directing the flow from the annular space to the center, e.g D.U.L.k. 131×110 mm.

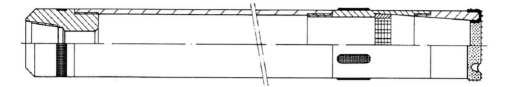

Figure 5: Single tube core barrel

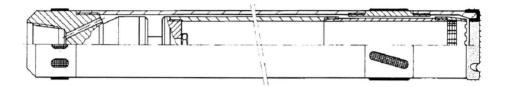

Figure 6: Double tube core barrel D, T-2, T-6

Figure 7: Double tube core barrel parts

3.3 Wire-line core barrel

Tube-in-tube system

A special design of the inner-tube allows to be retracted by means of a "catcher" fixed on a steel wire. Wire-line is much more comfortable, especially for deep drillholes, with the additional advantage to have the borehole cased. Core barrel and drill rods protect sensible rock sections from collapse. Having the inner tube removed, geophysical and hydrological measurements can be carried out within the borehole.

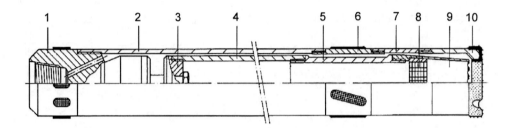

Figure 8: Double tube core barrel K-3 (heavy duty version, all parts threaded): 1 head, 2 outer tube, 3 inner tube head, 4 inner tube, 5 connecting sleeve, 6 reamer, 7 outer connecting sleeve, 8 core lifter, 9 core lifter case, 10 core bit

Outer tube system consists of bit, reamer, outer tube, landing sleeve and locking coupling.

Inner tube system consists of core lifter case incl. core lifter, stabilizer, inner tube, core barrel head incl. landing ring and latches with spring.

Wire-line core barrels are available in a wide range of sizes, starting mainly from 48 to 176 mm. International standards are:

- DCDMA sizes: AWL / BWL / NWL / HWL / PWL

- Metric sizes: NSK 146 / SK6L 146 / Geobor S / CSK 146 / CSK 176

Manufacturers produce according to these standards, but they also have own developments, e.g. for better core recovery, more sturdy design, modified catching systems.

Central european drillers prefer the bigger core size of 100 mm (Geobor S 146×102 mm / CSK 146×102), Austrians also use the CSK 176×132 to have the most heavy core with 132 mm, which means a weight of about 35 kg per meter.

Exploration drillings in ore, salt and coal mines use BWL (60×38 mm), NWL (75×47 mm), HWL (96×63 mm).

Looking at the dimensions of borehole and core, the bit for a wire-line core barrel has a very thick kerf compared to single tube, double tube core barrel: This thickness is needed to fix all these sophisticated parts and retract the inner tube.

The price for bit and barrel is much higher related to standard core equipments. But it is the only means to realize deep core drilling in a safe and economic way.

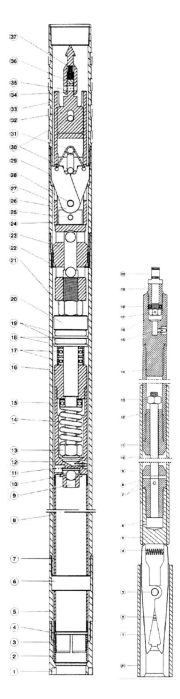

Figure 9: left: Complete core barrel assembly (outer tube / inner tube system); right: overshot (catcher for the inner tube)

(Note: The large number of parts give an idea of the complicated construction)

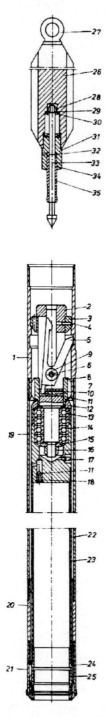

Figure 10: Wire-line core barrel SK6L (complete assembly)

Figure 11: CSK 146 core barrel head (part), overshot (catcher for the inner tube)

Figure 12: Geoline – Geobor S, dry hole device, core barrel head, overshot

Figure 13: SK6L overshot, core barrel head

Figure 14: Tungsten carbide bits

4 Flushing

The friction associated with the rotation of the bit produces heat, which burns the bit and, in particular, the diamonds. Therefore, flushing and mud control is of great importance for the bit lifetime, performance and core quality. The bit needs to be cooled, cuttings must be removed quickly from the bottom of the hole. Drill rods must be lubricated to reduce friction and wear. Using only clear water can erode the core and the borehole wall; sedimentation acts like cement to the rods. Flushing additives like bentonite, CMC, polymers help to get a better core recovery.

5   Bits, core bits

The most sensitive part in the complete core barrel system is the bit. It is the frontier tool, it faces the rock. The core bit must cut an annular space and creates the core. Quartz is widespread and abrasive. Other minerals and rocks can also be very hard, e.g. basalt, but are less abrasive. Diamond is the most suitable material to cut any kind of rock.

We summarize:

- The softer a rock is, the bigger should be the diamond grains

- The harder a rock is, the smaller should be the diamond grains

- Diamond "dust" in impregnated bits (for materials harder than granite)

- Small stones (40-60 spc[1]) surface set bits (for materials harder than basalt)

- medium stones (20-30 spc) for limestone

- big stones (6-10 spc) for salt and gypsum

- Synset / TSD elements (3 mm long) for limestone

- Stratacut / PCD elements (⌀ 12 mm) for arcose and marl

- TC-tungsten carbide hexagonal bolts, mainly applied to soft overburden soil

A skilled driller can calculate the best possible parameters for his bit, e.g.:

Geobor S bit 146×102 mm: 61 carats, 20/25 spc, 1500 diamond grains with 70% in the cutting front ($\sim$ 1000 grains) and 30% in the inner and outer gage. The received force is ca 3 kg load on each grain in the front, in total 3000 kg or 3 tons!

Typical drilling performances are:

very hard rock	2-3 cm/minute
medium hard rock	5-8 cm/minute
soft, incoherent rock	9-12 cm/minute

---

[1]stones per carat

Figure 15: TC and diamond bits for single tube, double tube and wire line core barrels

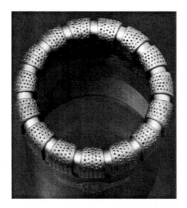

Figure 16: Surface set diamond bit

Figure 17: Synset-(Ballas) diamond bit

Figure 18: Stratacut diamond bit

Figure 19: Impregnated diamond bit

Appendix

The following abbreviations are used in this text:

DCDMA: Diamond Core Drillers Manufacturers Association

GOST: GOsudarstvennyi STandart (Russian Standard)

T, T-2, T-6, K-3, D: Metric standard for core barrels

modified D: D-core barrel with liner inside

D.U.L.k.: D-core barrel with Umlenkkopf (a head to press the core with a flush pump)

AWL, BWL, NWL, HWL, PWL: American sizes for core barrels and rods (WL: wire line)

NSK: Nassovia Seilkernrohr *(Nassovia wire line core barrel)*

SK6L: Seilkernrohr *(wire line core barrel)* 6" Large

Geobor S: Trademark of AtlasCopco company

CSK: Type of wire line core barrel with 5 latches in the head

TSD: Thermo stable diamond (synthetic diamond)

PCD: Polycristalline diamond

CMC: Mud additive, carboxyl-methyl-cellulose

References

[1] Happel, M.: Kernbohrausrüstungen und Spülbohrverfahren zur Bauwerkserkundung – Eine Auswahl gängiger Bohrsysteme, Comdrill Bohrausrüstungen GmbH, internal report

[2] Happel, M.: Von der Kernbohrung zur Injektion – Betrachtungen zur Probengewinnung und Einleitung von Maßnahmen zur Verbesserung des Zustandes von Bauwerk und Baugrund, Comdrill Bohrausrüstungen GmbH, internal report

[3] Heinz, W.F.: Diamond Drilling Handbook, 3rd edition, 1994 (can be ordered via email directly from the author: info@rodio.co.za)

[4] Drilling – the manual of methods, applications, and management, Australian Drilling Industry Training Committee Ltd. (ed.), 4th Edition, CRC-Press, 1997

[5] Nguyen, J.P. & Gabolde, G.: Drilling Data Handbook, Editions Technip, 1999

[6] COMDRILL Product Catalogue; COMDRILL Drilling Equipment GmbH 6th Edition 2004; Part 1: Diamond and TC-Tools, Core barrels, Casings, Drill rods, Tricone Bits, Drag bits, Down-the-Hole-Hammers, Sampling tools, Handling tools.

Tunnel support ahead of the face with pipe roofs or forepoling

Wolfgang Holzleitner, Franz Deisl, Wolfgang Holzer

Bernard+Partner ZT-Ges.m.b.H., Bahnhofstr. 19, A-6020 Hall in Tirol

Abstract: Conventional tunnelling in loose ground requires a support of the overhead excavation section ahead of the face. The support elements may consist of different kinds of forepoling or of a pipe roof. The latest developments of these can be installed by the drilling jumbo used for tunnelling works. The selection of the utilized ahead support elements, the workmanship, the influence on cost and time are dealt with as well as contractual and design aspects.

1 Introduction

This paper introduces and compares the various possibilities of applications of forepoling and pipe roof support in loose ground in order to analyze why the use of pipe roofs has increased considerably through the past few years. The described pipe roofs are installed with an equipment which can be mounted on the drilling jumbos and are available on the market as 'AT-Hüllrohr System', 'Rohrschirm System' or 'SYMMETRIX'.

This paper is based on the experience made during site supervision and site consultancy under consideration of geotechnical measurement results on recent tunnel projects on the northern approach to the Brenner Base Tunnel with a total number of 40,000 running meters of roof pipes and 250,000 running meters of forepoling spiles.

1.1 Review of the development

Until recent years, pipe roofs represented a heavy construction method. Thus the ahead-support with roof pipes was done with another crew and another equipment and as the tunnel excavation works. Hence, due to the alternating employment of tunnellers and pipe roof teams this method involved high cost, as one or the other team had to be on standby and exact prediction of cycle times in loose ground tunnelling is hardly possible. In order to guarantee progress without interruption, two teams had to be kept operational on site, see Fig. 1.

With the introduction of pipe roof systems which can be installed from drilling jumbos it became possible that pipe roof and tunnel excavation can be done by

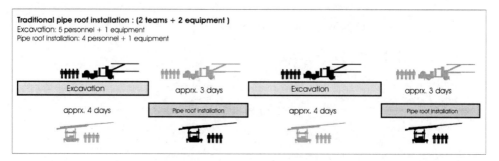

Figure 1: Traditional pipe roof installation with two teams

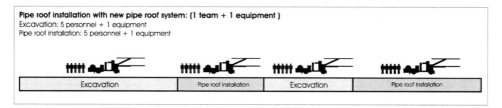

Figure 2: Pipe roof installation with new pipe roof system with one team

one and the same team of miners. This resulted in considerable cost reduction and the use of pipe roofs became a common standard measure for ahead-support, see Fig. 2. In conventional tunnelling in loose ground, forepoling and lagging sheets were the key to the success of any tunnelling project. The various types of forepoling have now been replaced to a considerable extent by the new pipe roof systems.

2   General observations

2.1   The action of 'ahead-support'

Tunnelling in loose ground is only possible due to the 3D arching and three-dimensional load transfer in longitudinal tunnel direction across the temporarily unsupported tunnel face. In order to reduce or avoid an overstressing of the arch and face, loose ground has to be reinforced by support ahead of the face. The support measures ahead of the face protect parts of the ground to drop. An excavation according to the theoretical profile induces shorter span and hence the displacements in the tunnel and on the surface become a minimum. The installation of ahead-support depends to a high extent on the quality of workmanship.

Figure 3: Pipe roof installation using a drilling jumbo

2.2 Scope of ahead-support

The scope of the ahead-support may be defined as:

- Achievement of the theoretical excavation profile

- Maximization of the advance step

- Immediate adjustment to the local ground properties

- Minimum time of production

- Minimum of cost.

The requirements to ahead-support appear to be the merging of contradictions: On the one hand, the ground requires improvement in order that it does not collapse during stress redistribution and on the other hand, the ground is disturbed during the installation of such ahead-support. Therefore the main target after installation of the steel elements is to compensate the disturbance which has been caused due to installation. In most of the designs it is assumed that an ahead-support improves the ground when combined with grouting. As can be seen along the excavation contour during excavation this target is often not achieved. Theory and practical aspects are somehow contradictional in this respect and only careful execution can achieve a convergence between theory and field application.

Figure 4: Overbreak above the spiles

3   Ahead-support with forepoling

3.1   Types of forepolings – Definitions

Spiles: Spiles are corrugated reinforcement bars with a diameter of 26 mm and a length of 3-4 m. They can be driven into fine to medium coarse, loose to medium dense layered grounds by a special adapter on a boom with a spacing of 20-30 cm. The ground in the vicinity of the spiles will be compacted and receives a type of 'prestressing'. Hence, spiles achieve good profile shapes.

Predrilled ungrouted spiles: Ungrouted spiles are made of corrugated reinforcement rods with a diameter of approximately 26 mm and a length of 3-4 m. They are installed in medium to coarse ground and are inserted in pre-drilled holes with a spacing of 20-30 cm. However, this type can hardly be used in loose ground, as drill holes collapse and cause loosening. Overbreaks above the installed spiles are the consequence. The high weight of the bars stands against a low stiffness compared to a forepoling pipe.

Grouted spiles: Grouted spiles can not be applied in loose ground, as no stable borehole for the insertion of the grout hose can be achieved.

Ungrouted, self-drilling forepoling: Self-drilling forepoling consist of hollow steel rods with a diameter of 25-38 mm and a length of 3 m or longer which can be extended by couplings via the external continuous thread and have a sacrificed drill bit. Self drilling bolts are installed by the boomer. They can be installed in medium to coarse ground with a spacing of 20-30 cm. Since the disturbance of the ground due to the installation of the rods is not compensated, overbreaks above the desired profile shapes may frequently occur, see Fig. 4.

Grouted self-drilling forepoling: The ideal type of installation is to use the grout as the flushing liquid. This type of installation warrants best compound between the ground and the forepoling rod. To use cement grout for flushing, the drilling jumbo has to be equipped with a special cooling system and an adapter between the hammer and the rod is required to convey the grout into the flushing duct. When using grout as the flushing aggregate the equipment is polluted with cement grout and therefore contractors are not really fond of this application. Since it delivers very good results, his type of installation should be propagated by designers and consultants.

The more frequent type of installation of grouted self-drilling forepoling takes place by drilling with air and/or water flushing with grouting as the succeeding working step. It is observed, that a complete filling of the borehole is hardly the case although a complete mortar embedment is desired. The reason is that the relatively thin grout will escape in a ground with coarse pores and the annular space of the borehole will remain empty and will cause loosening.

Principally, support elements such as forepoling or rockbolts which are installed upwards should be grouted upwards. The self-drilling forepoles are grouted downwards (from tip to toe) although they are installed upwards. Hence unsatisfactory grouting results do not surprise.

Ungrouted forepoling pipes: Forepoling pipes are steel pipes with a diameter of approximately 50 mm and of wall thickness between 3-5 mm with a welded on tip or with a skew open end. They are installed at lengths of 3-4 m in fine to medium grained grounds either driven or inserted in pre-drilled (partly collapsed) holes in a dense spacing of 20-30 cm. Due to the smaller diameter of the drill hole the ground will be compacted while the pipes are driven in. Hence a good profile shape can be achieved. The production of the drill holes should be adapted to the ground conditions either with air or water flushing.

Grouted forepoling pipes: Forepoling pipes of 3-4 m lengths with holes or slots for grouting purposes are installed like the above described ungrouted forepoling pipes and are grouted in the succeeding working step. Good compound is achieved between the grout and the ground and an exact profile shape is the result.

3.2   General rules for the execution of forepoling

The target of any advance support is to install the support elements into the ground ahead of the face in the shape of the excavation profile, so that the excavation profile can be achieved without major overbreak. Due to the installation of the ahead-support the ground experiences a certain disturbance which is to be compensated with appropriate grouting measures. The use of steel ribs is a precondition for the installation of the ahead-support, as the steel rib with shotcrete forms the support of the ahead-support at the place of the nearest rib to the face. The effectiveness of an ahead-support depends also on the length of advance steps; principally, ahead-support with forepoling poles or spiles is only effective with lengths of advance steps of less than 1.2 m in loose ground. When installing ahead-support, the shotcrete lining above the steel rib should be kept open for installation of forepoling and only when the ahead-support has been placed, the shotcrete should be applied in these places in order to achieve an appropriate support and embeddment for forpoling at the steel rib next to the face. Attention should be paid that forepoling is embedded in shotcrete also between the steel rib next to the face and the tunnel face, which will result in a minimum overbreak during the excavation of the next advance step, see Fig. 6.

As the grout mortar should have sufficient strength before starting the next excavation step, the use of accelerated cement is favourable; the use of accelerated cement increases the time for cleaning and wear, so that normal cement is prefered when possible, see Fig. 5.

In choosing the length of ahead-support it should be considered that the steel rods or pipes have sufficient support lengths in the ground and on the steel ribs.

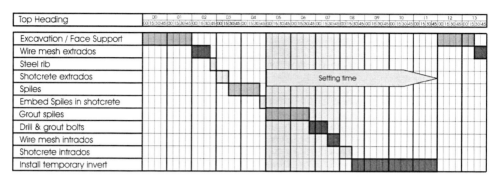

Figure 5: Excavation & support-cycle with grouted ahead-support

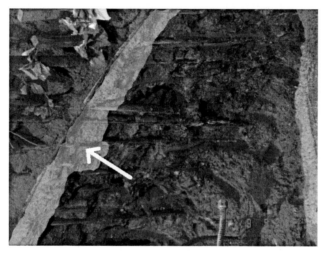

Figure 6: Shotcrete between steel rib and face (arrow) lead to good profil-shape during the upcomming excavation-step

Forepoling pipes should not have grout holes for approximately 1 m at their ends in order to avoid escape of the grout at the tunnel face. Prior to grouting, the mouth of the boreholes should be sufficiently sealed in order to achieve sufficient grouting embedment of the pipes. In order to enable a certain grouting pressure, a mechanical packer including a tap should be attached at every end of the pipes, which can be reused of course.

In loose ground, the flushing of self-drilling rods should be done by means of cement grout as this method warrants best results without any loosening effects.

Longer ahead-support with self-drilling bolts with a length of 6 m which may then be installed only every second advance step have turned out to be not ideal, as the material tends to overbreak at least as far as the installed rods; so this 'saw-tooth' profile requires a considerable amount of excess shotcrete, which is uneconomic.

3.3   The advantage of forepoling poles or pipes

A great advantage of the forepoling poles or pipes is that the various types and lengths can be adjusted to the encountered loose ground properties. Different types of rods within the same advance step such as grouted forepoling pipes in the area of coarse gravel-like formations while the other parts of the tunnel wall may be supported with spiles in the finer formation. A further advantage is that it can be immediately responded upon a change of ground properties; this may be an omission of support in certain sections or an increase in number and length or change of type.

3.4   Disadvantages of forepoling pipes or poles

Disadvantages of these types of ahead-support is the limited length of 3-4 m which may result in a lower theoretical safety at the tunnel face, however observations of the overbreaking tendency of the tunnel face and immediate support with shotcrete can counteract this shortcoming.

A further disadvantage compared to pipe roofs is that no information about the type of material occuring the next meters of tunnel excavation can be obtained.

4   Ahead-support with pipe roofs

When installing a pipe roof, it is required to enlarge the profile for approximately 20% in cross sectional area in order to be able to install the pipe roof above the required excavation profile. No constant but a variable profile in the shape of a cone is required, see Fig. 7; the length of this geometry depends on the length of one excavation length until the next pipe roof is installed; such lengths vary between 10-15 m, the excess length amounts to 3-4 m. When installing the inner lining the varying thicknesses are unfavourable so that the overprofiles should be smoothened by shotcrete or pre-lining concrete.

4.1   Selection of pipe roof pipe diameter

Due to the absence of dimensioning rules, the selection of the pipe diameter (76 mm/114 mm/139 mm) is limited to experience. Calculations can only

Figure 7: Pipe roofs cause 'saw-tooth'-shaped tunnel excavation

roughly support the decisions. Due to flushing during the drilling of the pipe roof pipes, a considerable disturbance of the ground may occur. The larger the pipe diameter the more probable is this disturbance. At shallow overburden large pipe roof pipe diameters may result in measurable surface settlements during the phase of the pipe roof installation. This disadvantage can be counteracted by installing every second pipe with immediate grouting first and in the second working step the missing pipes can be installed. The accessibility to the second pipes can be achieved by cutting off the ends of the already installed pipes. Attention should be paid that the ground between the pipes does not collapse. The minimisation of the spacing of the pipes may represent a counterproductive measure. The production time of pipe roofs is significantly influenced by the pipe diameter, which means that smaller diameters can be installed in much shorter periods. The material is better flushed out when using larger diameters. When using pipe roofs of 15-20 m, the use of diameter 114 mm appears to be a good compromise. When using pipe roof pipes for the exploration drillings, a more detailed prediction can be made with the larger diameters, also the achievable depths are longer.

4.2 Features of pipe roof pipes

Pipe roof pipes without valves: In ground of low water sensitivity the production of the pipe roof can proceed without valves. A condition for this is, that the loss of flushing water does not result in clogged casing pipes.

Pipe roof pipes with valves: In erosive ground the migration of flushing water into the ground should be omitted as the borehole walls can be eroded and the ground between the pipes may collapse. The use of valves is also required when a complete loss of flushing water is observed.

Pipe roof pipes with internal steel rods: When expecting high loads

immediately after excavation, pipe roof pipes may be furnished with
reinforcement rods in order to achieve additional tensile strength in
particular at the thread connections. The internal steel rod is placed
with spacers in the pipe roof pipes and is grouted fully with cement
grout and represents a post-failure safety.

4.3   Types of grouting

Prime target of the grouting is the compensation of the ground disturbance
caused during installation. Further, a kind of prestressing of the ground
between the pipes would be favourable; however this is very seldom the case.

Pipe roof pipe without valves: The pipes are filled with cement grout by
   means of an attached cap at the mouth of the borehole; through the
   large opening at the end of the pipe at the far end of the borehole the
   majority of the grout will disappear, and in the annular space no cement
   grout can be found. This can be overcome by closing the far end of the
   pipe by means of some plastic tap or sponge as used when cleaning
   concrete pumps.

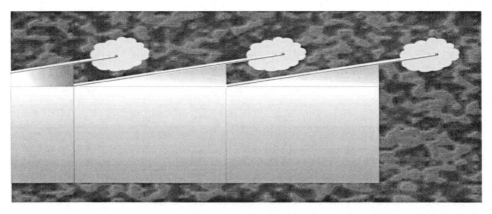

Figure 8: Assumed result of pipe roof grouting

Pipe roof pipe with valves: In case that a grouting section by section
   with a double packer is intended, the pipes are to be equipped with
   valves; without valves the grout would short-circuit into the pipe. When
   grouting through the valves, only the use of a double packer makes
   sense, since single packers cause a large waste of grout and outflow at
   the mouth of the pipe; in such cases the packer could only be moved
   with the drill rig due to the high weight of the grout column. After
   grouting section by section the pipe can be closed with a cap at the

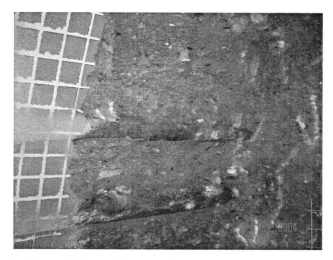

Figure 9: Exposed pipe roof pipe without grout traces

mouth of the pipe and filled up with grout; this increases its stiffness. A ground improvement which is comparable with a 'tube in manchette grouting' is not possible since the valve is not able to fulfil the cracking of the cement bentonite mixture in the annular space between pipe and borehole wall. The valves of the pipe roof pipes can only be used once, that means that several phases of grouting is impossible. It was also observed, that the valves are sometimes lost by vibration when drilling through boulders or cobblestones; the loss of such valves is noticed during grouting when short-circuits and escape of grout at the mouth of the pipes occur.

4.4 Advantages of pipe roofs

The advantage of ahead-support pipe roofs is the longer embeddment length ahead of the tunnel face; it causes a better load distribution from the face area into the ground. During the drilling of the pipes records can be made about drilling advance, colour of flushing water, the type and shape of aggregate and water inflows at the time of the extension of the pipe. Useful additional information can so be gained about the ground conditions within the next section of advance, see Fig. 10.

Roof pipes are also very well suitable as drainage pipes, see Fig. 11. It should be considered that such drainage pipes are not installed at the tunnel face since the water from the pipe influences the stability of the tunnel face. Drainage pipes are preferably installed from the side walls at sharp angles forward towards the driving direction; further it should be taken into account that such pipes can only be drilled slightly upwards – never downwards.

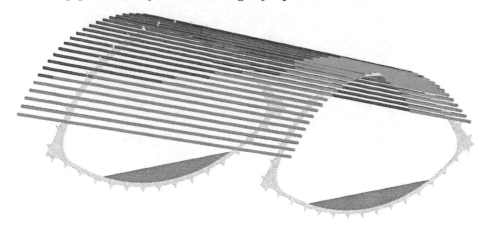

Figure 10: Exploration-results based on drilling records

Figure 11: Drainage drillings ahead of invert

4.5   Disadvantages of pipe roofs

One disadvantage of the pipe roof is that the ground is disturbed during the installation of the pipes. Ground loss by flushing can be higher than the volume of the pipes which, in consequence will result in surface settlements. Ground water may further be directed into the area of excavation as the vicinity of the pipes will act as a percolation path. Pipes which deviate from their theoretical line into the excavation profile are difficult to remove since they are only accessible in the unsupported area. Cutting off is to be done by disk grinder, since removal with the excavator would cause a further disturbance of the ground. Diverted pipes, when removed leave a gap in the ahead-support which has to be compensated by additional forepoling.

A further disadvantage is that before installation of a pipe roof the profile has to be enlarged for approximately 20% in order to create a headwall where the pipe roof can commence; this requires a certain length of transition in order to be able to position the boomer at the correct geometry of the conical shape of the pipe roof.

Due to this particular geometry which repeats every 10-15 m a 'saw tooth' is created which causes an excess use of inner lining concrete. Also the frequent change of thickness of the inner lining is unfavourable for the quality of the concrete.

In most cases, the rate of advance when using pipe roofs is lower than when using forepoling as the installation of a pipe roof consumes 36-48 hours at least. This results in higher time depending cost.

5   Contractual aspects

5.1   Who bears the risk of overbreak

The correct workmanship, which, as demonstrated above, plays an important role in loose ground tunnelling, is the responsibility of the contractor. The employer, when accepting the ahead-support for payment should not bear the cost of overbreak in addition. Exceptions due to unforeseen geological conditions are possible. Clear contractual regulations concerning the risk of workmanship will result in appropriate quality.

5.2   Ahead-support in the design drawings

When following the philosophy of the New Austrian Tunnelling Method that support measures should be installed depending on the conditions of ground or the permissible displacements, the design has to state the minimum of required tunnel support especially in cases when ahead-support such as pipe roof are incorporated into the structural calculation of the surface settlements.

6   Summarizing comparisons

6.1   Length ahead of the face

The smaller embeddment lengths of the forepoling poles or pipes ahead of the face has been stated as a disadvantage; however it should be noted that at the end of any pipe roof the embeddment lengths ahead of the face in case of pipe roofs are of the same amount as with forepolings even that the cross-section of the profile at the end of the pipe roof is approximately 20% larger. A support of the tunnel face which follows the rules of workmanship as seen in Fig. 13, consisting of shotcrete, wire mesh and long face anchors (15-20 m) and the excavation in subdivision as well as a support body, will warrant a reliable support of the forepoling pipes ahead of the face.

Excavation under the protection of pipe roof with much longer embeddment lengths ahead of the face should seemingly be able to proceed at longer excavation lengths as under forepoling; however in most of the cases this is not possible since the extension of the lengths of advance step will cause dropping ground in side walls and roof. Hence the excavation lengths are more or less the same (approximately 1.0 m) when under a pipe roof or forepoling, which means that the higher cost of pipe roof can hardly be compensated by increasing advance rates. The higher cost stand against an unquantifiable increase of safety.

6.2   Excavation cross-sections under pipe roofs

The 'saw tooth'-like shape caused by installation constraints cause an excess of excavation quantities, support quantities and inner lining concrete quantities. Pipe roofs are generally used in more difficult ground conditions than the forepoling; this stands in opposition to the fact that larger cross-sections cause more displacement in the tunnel and particularly above ground.

6.3   Omission and reinstatement of pipe roofs

As already explained, a pipe roof always requires a head wall for installation which repeats after every 10-15 m depending on the length of the roof pipes. When a tunnel is excavated under forepoling pipes and it is decided that a pipe roof is to be applied, an enlargement has to be made under the protection of forepoling; due to constraints of the drill rig and the geometry, the length of such a transition should be at least 6-8 m; this means that an immediate installation of the pipe roof is not possible (Fig. 12). Switching from pipe roof to forepoling is never a problem as the forepoling can always be installed from the cylindrical section.

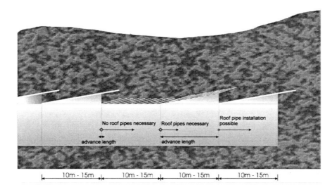

Figure 12: Switching between forepoling and pipe roofs

Figure 13: Careful excavation and support of subdivided tunnel face

6.4 Forepoling and the influence on excavation costs

Since the density of required ahead-support is directly linked with the ground conditions and hence with the responsibility of the employer, the ahead-support quantities should be paid by him. As price analysis of tunnel-bids shows, the used density of ahead-support has considerable influenced the excavation costs. Fig. 14 shows a variation of approximately 50% of the excavation costs. As forepoling poles or pipes can be varied in density as required by the ground conditions, the cost saving potential is high, whereas with pipe roofs which cannot be varied with the requirements of the ground cause higher cost; in the analysis reflected in Fig. 14 this amounts to 164%, which is a difference of 30% between an average forepoling class compared to a pipe roof class.

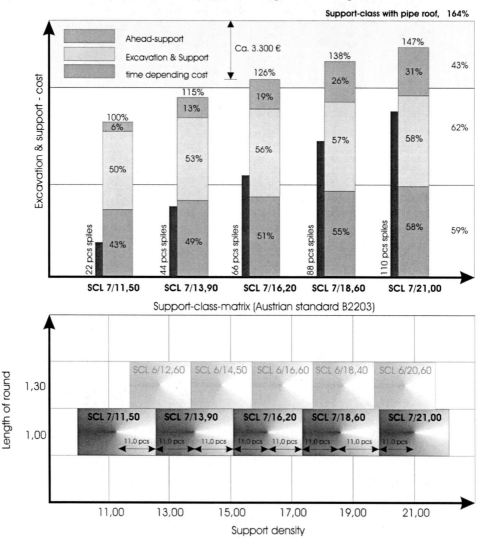

Figure 14: Cost comparison between forepoling and pipe roofs

6.5   Comparison of advance rates

In the more advanced tunnelling contract, the time depending cost of the site are paid in conjunction with the encountered rock classes; this means that every bid has to contain advance rates for every type of support class. Fig. 15 shows on the time location diagram the range of advance rates for tunnelling under forepoling and under pipe roofs including the period of installations of the ahead-support. This picture shows that overall advance rates of drive with forepoling-ahead-support as somewhat 40% higher as drive under pipe roofs.

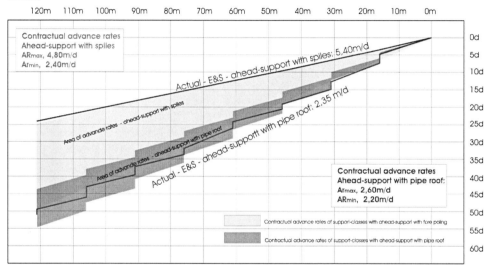

Figure 15: Comparison of advance rates

7   Summary

After an elaborate analysis of the observations and experiences with both types of aheadsupport, it is concluded, that both pipe roofs and forepoling are appropriate and technically comparable means of support ahead of the face.

The flexible adjustability of forepoling to the encountered ground conditions coincides with the philosophy of NATM that the ground should be improved to an extent in order to withstand the additional strains due to the stress flow redistribution.

An appropriate workmanship and the continuous working cycle with well trained tunnelling crews is a precondition for a successful application of loose ground tunnelling in the conventional method.

The equipment for the installation of pipe roof pipes can also be used for drainage and exploratory purposes so that the availability of such attachments on the drilling jumbo is a big advantage for any loose ground tunnelling site.

The disadvantages of the pipe roofs in view of advance rates and cost are obvious but did not influence the technical conclusion of this comparison.

New approaches in tunnelling with composite canopies – installation-design-monitoring

Ernst F. Ischebeck

Ischebeck GmbH, D-58242 Ennepetal, Germany

Abstract: The umbrella arch method of support, so-called forepoling, spiling or pipe roofing, is the most commonly used method since 20 years to improve poor ground conditions ahead of the tunnel excavation. Since some years direct drilled and roto-injected IBO-micropiles according to EN 14199 are an alternative and are successfully used in tunnelling in soft soils, where boreholes collapse.

The loading mechanism of micropiles is characterized by axial tension or compression of the reinforced groutbody, whereas pipe-roofs are loaded by bending.

Canopies TITAN are installed by the common crew with rocket boomer for approximately 2/3 of the costs of pipe roofing. Cement grout flushing of lightweight micropiles helps to stabilize the tunnel face instead of reduction of cohesion by installing pipe roofing.

The design of composite canopies is based on FE modelling or simple statics. Length, inclination and density of micropiles can be adapted to site conditions. Exhumed micropiles confirm deviations below $\pm 1\%$ of length of micropile. Composite canopies require a cohesion of soil $c \geq 50$ kN/m^2.

1 Introduction to micropiles

Micropiles are directly drilled and grouted by roto-injection (jet-grouting) in one visit with a sacrified drill bit.

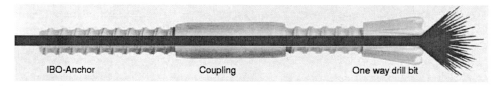

Figure 1: TITAN micropile

The immediate grouting improves the cohesion of soil and creates a composite section with a very stiff shear bond between steel and groutbody, which can be compared with other composite materials like reinforced concrete. The installation of micropiles TITAN can be adapted to nearly all tunnel construction sites.

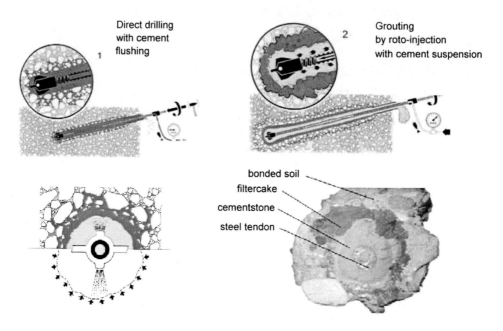

Figure 2: Installation of anchorpiles TITAN

To avoid axial cracks a minimum cement stone cover of 30-50 mm is necessary. The diameter D of the drill bit equals ca $3d$, where d is the diameter of the tendon. The diameter D^* of the grouted soil column equals approximately $(3 \ldots 4)d$.

2 Design of composite canopies

The tunnel advance e without support is naturally limited by cohesion of soil c, specific weight of soil γ, friction angle φ and tunnel diameter D. The idea of composite canopies is to widen the tunnel advance e by reinforcing the soil.

Three basic bearing mechanisms are known to support the tunnel advance:

1. Bending resistance of stiff steel tubes, forming additional an arch transversal to the tunnel direction.

2. Increased cohesion by reinforcing with composite micropiles, which additional form an arch transversal to the tunnel direction.

3. Jet grouted horizontal columns, which form an arch transversal to the tunnel direction.

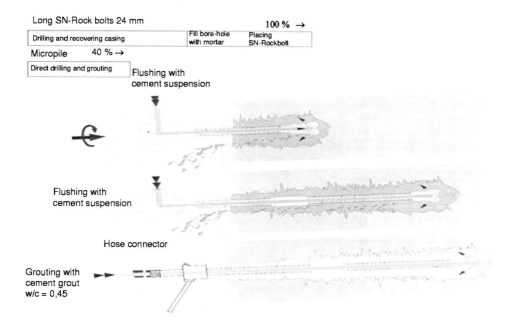

Long SN-Rock bolts 24 mm

100 % →

| Drilling and recovering casing | Fill bore-hole with mortar | Placing SN-Rockbolt |

Micropile   40 % →

| Direct drilling and grouting | Flushing with cement suspension |

Flushing with
cement suspension

Hose connector

Grouting with
cement grout
w/c = 0,45

Figure 3: Installation of micropiles

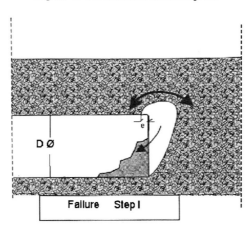

D Ø

e

Failure   Step I

Figure 4: Failure of tunnel face (step 1)

Bearing mechanism No. 2 by increased cohesion works as a reinforced concrete slab, as a hammock or a rope in a suspension bridge. Each micropile is loaded only by tension or compression forces. Each micropile is bonded at one end ahead of the tunnel face, or better aheat of the slip circle, e.g. > 4 m and at the other end to the stiff shotcrete tunnel lining.

For static calculation and design of composite canopies the loading mechanism can be modelled by:

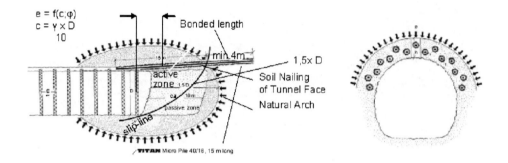

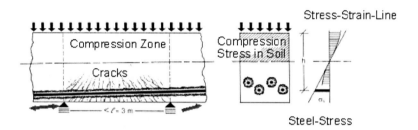

Figure 5: Loading and failure mechanism for composite canopies

- soil nailing and rope static (Fig. 6)

- silo effect (Fig. 7)

- FEM (Fig. 8)

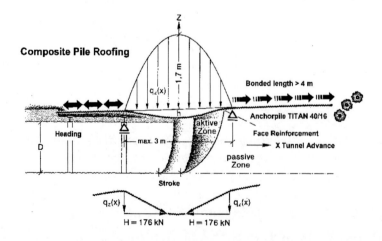

Figure 6: Composite pile roofing

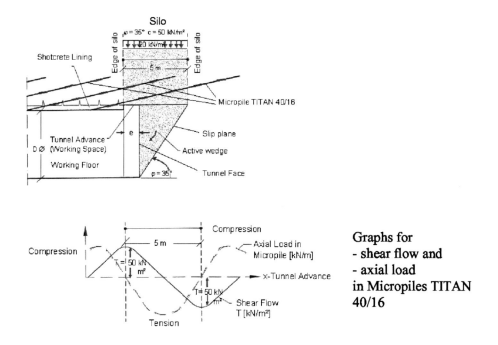

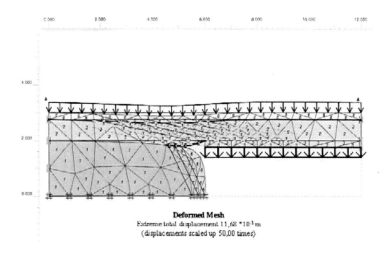

Figure 7: Distribution along the tunnel axis

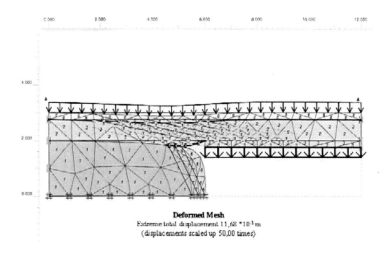

Figure 8: Modelling by PLAXIS [3]

As a result of FEM-modelling (FLAC-3D) Fig. 9 shows that stiffness of support by steel pipes 140 mm ∅ × 8 mm is similar to composite micropiles TITAN 73/53. Comparable settlements are estimated for both technologies.

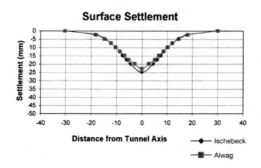

Figure 9: Surface settlement

3   Monitoring and site experiences

In 2001 the first successful tunnel drive with composite canopies was the access tunnel ARA[1] in the region Bern/Switzerland. The canopies were designed with PLAXIS 2D. The settlements were monitored by FM Messtechnik AG in Bern. The access tunnel was excavated in sand/gravel and supported with micropiles TITAN 40/16, very close underneath a waste water treatment plant.

Another tunnel drive with composite canopies was the Metro Santiago access tunnel "Pique Estacion El Parron" in 2002. The monitoring confirmed the estimated settlements.

The first application in Germany was in 2004, tunnel Berghofen/Dortmund in schist and weathered claystone.

---

[1] Abwasserreinigungsanlage: sewage treatment plant

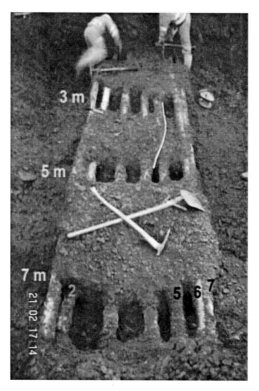

Figure 10: Exhumed micropiles TITAN 40/16 of a composite canopy

$D = 100$ mm, weathered mud stone measured deviations $< \pm 1\%$ of length up to 12 m

References

[1] Twenty years of experiences in the use of the umbrella-arch method of support for tunnelling; G. Carrieri, Geodata; R. Fiorotto, Casagrande; P. Grasso, Geodata; S. Pelizza, Politecnico Torino; IWM – Micropiles Venice, 30 May - 2 June 2002 in Venice

[2] Möhrke, Tunnelvortrieb an der Eisenbahnstrecke Platamon-Leptokaria (Tunnel drive at the railway line Platamon-Leptokaria), *Felsbau* 17 (1999) Nr. 5

[3] Erdstatische Berechnung, Spiessvortrieb mit Injektionsankern TITAN 30/11, ARA Region Bern AG, Ingenieurbüro Dr.-Ing. Ludwig W. Zangl, 28.8.2001

[4] Setzungsmessungen Zugangsstollen (First), FM Messtechnik AG, Bern, 6.3.2002

[5] Design of Ischebeck pipe roof system for the Metro Santiego, "Tunnel de Acceso Pique Estacion El Parron", Geoconsult ZT GmbH, Salzburg, 11.4.2002

[6] Prof. Dr.-Ing. B. Maidl: Verbundschirm aus IBO-Ankern Tunnel B 236 in Dortmund-Berghofen, baubegleitende Messungen, 2004-2005

Lighting of road tunnels in Austria

Rudolf Koller

Hopferwieser Consult ZT GmbH, Santnergasse 61, A-5020 Salzburg, Austria

Abstract: This paper gives a short guidance how to design a tunnel lighting system. It shows why artificial lighting in a tunnel is necessary, informs which national and international regulations are existing and gives the main lighting definitions. In the paper the main inputs are given, which are the basis of the lighting design and shows how the lighting shall be measured after installation and controlled.

1 Introduction

The lighting requirements of a tunnel are totally different by day and by night. At night the problem is relatively simple and consists in providing luminance levels on lit routes inside the tunnel at least equal to those outside the tunnel. The design of the lighting during day-time is particularly critical because of the human visual system. The driver outside the tunnel cannot simultaneously perceive details on the road under lighting levels existing in a highly illuminated exterior and a relatively dark interior (i.e. transient adaptation).

While the visual system can adapt to rapid reduction in ambient illumination, such as that produced when passing from daylight into the darkness of a tunnel, these adjustments are not instantaneous. The adaptation process takes a certain time, depending on the amplitude of the reduction: the greater the difference, the longer the adaptation time.

For a given speed, this means that the greater the difference between the lighting level outside and that inside the tunnel, the longer will be the distance over which the visual system of the driver has to adapt.

The requirements for lighting installation of a tunnel are influenced by several critical factors, which determine visibility. These conditions are eminently variable, and involve characteristics of the driver, including ability, age and personal habits; the physical conditions of the road, access to and length of the tunnel; atmospheric conditions; traffic density, volume and speed; and type of vehicles in transit. Additional considerations include the contribution of lighting to the architectural aspect of the tunnel portal with regard to visual guidance, comfort and to the overall maintenance of the installation.

As is the case for all lighting installations, the quality of tunnel lighting can vary as a function of some parameters. The minimum day-time and night-time lighting requirement is to ensure visibility conditions such that the user may travel through equally well by day and by night at a given design speed. It should provide safety, comfort and confidence at a level not lower than that which exists at the same time along the access roads to the tunnel.

In order to achieve this purpose, it is essential for road users to have, inside the tunnel, sufficient visual information regarding the geometry of the portion of the road forming the field of view, and the presence and movement of possible obstacles, the latter comprising particularly other road users. However, it is also necessary that motorists approaching the entrance of the tunnel should have the same feeling of confidence that they had along the preceding portion of the access road to the entrance.

The photometric characteristics of the lighting installation of a tunnel which define the quality of the lighting system are:

- The luminance level of the road and of the lower part of the tunnel walls;

- The uniformity and distribution of the luminance of the road surface and of the walls;

- The limitation of glare produced by the light sources;

- The limitation of the flicker effect;

- The level of visibility of possible obstacles;

- The visual guidance.

All of the values specified later on are values to be maintained throughout the duration of operation of the tunnel. In order to obtain the values to be achieved to the tunnel in the brand new state it is therefore necessary to increase the specified values to take the conditions of maintenance of the installation into account. They depend on the quality of the equipment used, the frequency of maintenance and the ambient conditions of the site.

2   Guidelines and regulations – national and international

Austrian regulation

- RVS 9.27 – Projektierungsrichtlinien Beleuchtung, Ausgabe Oktober 1991

- RVS 9.27 – Projektierungsrichtlinien Beleuchtung, Entwurf November 2004 (Version 12)

German regulation

- Richtlinien für die Ausstattung und den Betrieb von Straßentunneln RABT, Ausgabe 2003

- DIN 67524, Mai 1987 Beleuchtung von Straßentunnels und Unterführungen

Swiss regulation

- Leitsätze der Schweizerischen Lichttechnischen Gesellschaft, Ausgabe 2002

International guidelines and regulations

- International Commission on Illumination: CIE 88:2004, 2nd edition, guide for the lighting of road tunnels and underpasses

- CEN Report CR 14380, April 2003 - Lighting applications – Tunnel lighting

3 Definitions

3.1 Tunnel related zones

3.1.1 Entrance portal

The entrance portal is the part of the tunnel construction that corresponds to the beginning of the covered part of the tunnel.

3.1.2 Exit portal

The exit portal is the end of the covered part of the tunnel.

3.1.3   Access zone

The access zone is the part of the open road immediately outside (in front of) the entrance portal, covering the distance over which an approaching driver should be able to see into the tunnel. The access zone begins at the stopping distance point ahead of the entrance portal and ends at the entrance portal.

3.1.4   Threshold zone

Threshold zone is the first part of the tunnel, directly after the entrance portal. The threshold zone begins at the entrance portal.

3.1.5   Transition zone

Transition zone is the part of the tunnel following directly after the threshold zone. The transition zone stretches from the end of the threshold zone to the beginning of the interior zone. In the transition zone, the lighting level is decreased from the level at the end of the threshold zone to the level of the interior zone.

3.1.6   Entrance zone

The entrance zone is the combination of the threshold zone and transition zone.

3.1.7   Interior zone

The interior zone is the part of the tunnel following directly after the transition zone. The interior zone stretches from the end of the transition zone to the beginning of the exit zone.

3.1.8   Exit zone

The exit zone is the part of the tunnel where, during day-time the vision of driver approaching the exit is influenced predominantly by the brightness outside the tunnel. The exit zone stretches from the end of the interior zone to the exit portal of the tunnel.

3.1.9 Parting zone

The parting zone is the first part of the open road directly after the exit portal of the tunnel. The parting zone is not part of the tunnel, but it is closely related to the tunnel lighting. The parting zone begins at the exit portal.

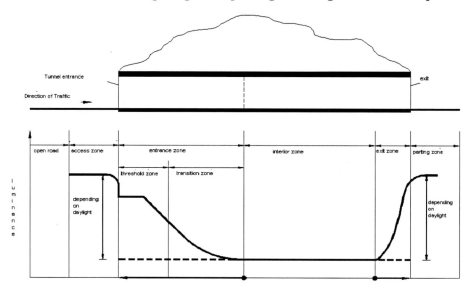

Figure 1: Schematic distribution of luminance during day-time

3.2 Lighting definitions

3.2.1 Contrast C

The contrast between a relatively small object with sharp contours and its (immediate) background is generally defined as:

$$C(\%) = 100 \cdot (L_o - L_b)/L_b$$

L_o: Luminance of the object
L_b: Luminance of the background

3.2.2 Luminous flux

The luminous flux is the radiant flux which is emitted from a light source. The evaluation is done using the spectral luminous efficiency curve $V(\lambda)$.

3.2.3   Luminance L (cd/m^2)

$$L = \mathrm{d}\Phi/(\mathrm{d}A \cdot \mathrm{d}\Omega)$$

L : Luminance (cd/m^2)
dΦ: Luminous flux fraction
dA: Surface entity
dΩ: Solid angle

3.2.4   Average luminance (cd/m^2)

The average luminance is the arithmetic mean of grid points on a particular section in the tunnel.

3.2.5   Illuminance E (lx)

$$E = \mathrm{d}\Phi/\mathrm{d}A$$

E : Illuminance (lx)
dΦ: Luminous flux fraction
dA: Surface entity

3.2.6   Luminous efficiency (lm/W)

The luminous efficiency represents the quotient from the emitted luminous flux and the electrical energy and the energy consumption of the radiation source (without fittings as ballasts).

3.2.7   Flicker

Flicker sensations are seen when driving through spatially periodic changes in luminance, such as those produced by daylight screens (both sun-tight and nonsun- tight) or luminaires that are mounted separately. Under specific conditions, the flicker may cause discomfort that sometimes can be severe.

3.2.8   Overall uniformity U_o

$$U_o = L_{\min}/L_{\mathrm{av}}$$

U_o : Overall uniformity
$L_{\min}$: Minimum road luminance
L_{av} : Average road luminance

3.2.9 Longitudinal uniformity U_l

$$U_l = L_{\min}/L_{\max}$$

U_l : Longitudinal uniformity in the centre of one lane
$L_{\min}$: Minimum road luminance
$L_{\max}$: Maximum road luminance

3.2.10 Reflection factor

The reflection factor of road surface is defined by reflection tables (class R1 to R4 resp. C1 to C2 with average luminance coefficient).

For different road surfaces and tunnel walls shall be used the following luminance coefficient:

Concrete surface	R2	$q_0 = 0.08$ cd/m^2/lx
Asphalt surface	R3	$q_0 = 0.07$ cd/m^2/lx
Tunnel wall		$q_0 = 0.10$ cd/m^2/lx

(q_0: mean luminance coefficient)

3.3 Tunnel specific definitions

3.3.1 Threshold zone lighting

Lighting of the threshold zone of the tunnel, which allows drivers to see into the tunnel whilst in the access zone.

3.3.2 Transition zone lighting

Lighting of the transition zone, which facilitates the driver's visual adaptation to the lower level in the interior zone.

3.3.3 Interior zone lighting

Lighting of the interior zone of the tunnel, which provides adequate visibility in the interior of the tunnel, irrespective of the use of vehicle headlights.

3.3.4 Exit zone lighting

Lighting of the exit zone, which improves the visual performance during the transition from the interior zone to the open road beyond the tunnel.

3.3.5   Vertical illuminance (E_v)

The illuminance at a particular location at a height of normally 0.1 m above the road surface, in a plane facing the direction of oncoming traffic. The height of 0.1 m above the road surface is meant to represent the centre of an object of 0.2 m × 0.2 m.

3.3.6   Visual guidance

The optical and geometrical means that ensure that motorists are given adequate information on the course of the road in the tunnel.

3.3.7   Emergency lighting

Lighting provided for use when the supply to the normal lighting fails.

3.3.8   Fire emergency guidance lighting

Lighting providing visual guidance in the event of fire and smoke.

3.3.9   Access zone luminance

The eye adaptation luminance in the access zone.

3.3.10 L_{20} access luminance

The luminance L_{20} in the access zone is defined as the average of the luminance values measured in a conical field of view, subtending an angle of 20° (2 × 10°), by an observer located at the reference point and looking towards a centered point at a height equal to one quarter of the height of the tunnel opening.

This average luminance is conventionally considered as representative of the state of adaptation of the eye of a driver approaching the entrance of the tunnel when he finds himself at the reference point and is used as a basis for computing the luminance in the entrance zone. Preferably, it can be calculated or it can be measured by means of a luminance meter having a 20° angle of aperture.

3.3.11 Equivalent ocular veiling luminance (L_{seq})

The light veil as a result of the ocular scatter L_{seq} is quantified as a luminance.

3.3.12 Atmospheric luminance (L_{atm})

The light veil as a result of the scatter in the atmosphere expressed as a luminance.

3.3.13 Windscreen luminance (L_{winds})

The light veil as a result of the scatter in the vehicle windscreen expressed as a luminance.

3.3.14 Threshold zone luminance (L_{th})

The average road surface luminance of a transverse strip at a given location in the threshold zone of the tunnel (as a function of the measurement grid).

3.3.15 Transition zone luminance (L_{tr})

The average road surface luminance of a transverse strip at a given location in the transition zone of the tunnel (as a function of the measurement grid).

3.3.16 Interior zone luminance (L_{in})

The average road surface luminance of a transverse strip at a given location in the interior zone of the tunnel (as a function of the measurement grid).

3.3.17 Contrast revealing coefficient (q_c)

The quotient between the luminance of the road surface, and the vertical illuminance E_v at that point

$$q_c = L/E_v$$

where q_c is the contrast revealing coefficient in cd/m^2·lx.

3.3.18   Threshold zone luminance ratio (k) at a point

The ratio between the threshold zone luminance L_{th} and the access zone luminance L_{20}:

$$k = L_{th}/L_{20}$$

3.3.19   Veiling luminance (L_v)

The overall luminance veil consisting of the contribution of the transient adaptation, the stray light in the optical media, in the atmosphere and in the vehicle windscreen.

3.4   Traffic related definitions

3.4.1   Carriageway

That part of the road normally used by vehicular traffic.

3.4.2   Traffic lane

A strip of carriageway intended to accommodate a single line of moving vehicles.

3.4.3   Emergency lane (hard shoulder)

A lane parallel to the traffic lane(s), not destined for normal traffic, but for emergency and police vehicles and/or for broken-down vehicles.

3.4.4   Traffic flow

The number of vehicles passing a specific point in a stated time in stated direction(s). In tunnel design, peak hour traffic, vehicles per hour per lane, will be used.

3.4.5   Speed limit

The maximum legally allowed speed.

3.4.6 Design speed

A speed adapted for a particular stated purpose in designing a road.

3.4.7 Reaction time

The minimum time interval between the occurrence of an event demanding immediate action by the driver and his response. The reaction time includes the time needed for perception, taking a decision and acting.

3.4.8 Stopping distance "SD"

The stopping distance "SD" is the distance needed to bring a vehicle, driving at design speed, to a complete standstill. The "SD" is usually defined in national legislation or regulation. The concept safe stopping distance is not used herein.

3.4.9 Mixed traffic

Traffic that consists of motor vehicles, cyclists, pedestrians etc.

3.4.10 Motor traffic

Traffic that consists of motorized vehicles only. It depends on national legislation which vehicle types are included in this classification. In some countries it only includes vehicles, which are capable of maintaining a minimum speed. In others, mopeds are not considered as motorized traffic.

3.5 Lighting systems

3.5.1 Symmetric lighting

The lighting where the light equally falls on objects in directions with and against the traffic. Symmetric lighting is characterized by using luminaires that show a luminous intensity distribution that is symmetric in relation to the plane normal to the direction of the traffic.

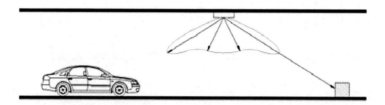

Figure 2: Scheme symmetric lighting

3.5.2 Counter-beam lighting (CBL)

The lighting where the light falls on objects from an opposite direction to
the traffic. Counter-beam lighting is characterized by using luminaires that
show a luminous intensity distribution that is asymmetric in relation to the
plane normal to the direction of the traffic, where the maximum luminous
intensity is aimed against the direction of the traffic. The term refers only to
the direction of normal travel.

Figure 3: Scheme counter-beam lighting

3.5.3 Pro-beam lighting

The lighting where the light falls on objects in the same direction as the
traffic. Pro-beam lighting is characterized by using luminaires that show a
luminous intensity distribution that is asymmetric in relation to the 90/270
C-plane (the plane normal to the direction of the traffic), where the maximum
luminous intensity is aimed in the same direction as the direction of the traffic.

3.5.4 Reference obstacle

Cube with a 0.2 m side and diffusing faces with a specified reflection factor
equal to 0.2.

4 Tunnel lighting design

4.1 Introduction

The requirements for lighting installation of a tunnel are influenced by several critical factors, which determine visibility. These conditions are eminently variable, and involve characteristics of the driver, including ability, age and personal habits; the physical conditions of the road, access to and the length of the tunnel; atmospheric conditions; traffic density, volume and speed; and type of vehicles in transit. Additional considerations include the contribution of lighting to the architectural aspect of the tunnel opening with regard to visual guidance, comfort and to the overall maintenance of the installation.

If, for special reasons, the tunnel must be completed with particular care, stricter requirements may be imposed and some information has to be given explicitly.

The minimum day-time and night-time lighting requirement is to ensure visibility conditions such that the user may travel through equally well by day and by night at a given design speed. It should provide safety, comfort and confidence at a level not lower than those that exist at the same time along the access roads to that tunnel.

4.2 Configuration of traffic

- Cars and trucks,

- bicycles and motorized bicycles,

- pedestrians.

4.3 Stopping distance

Important parameters for the design of tunnel lighting installations include the speed, volume and composition of traffic flow entering, and passing through the tunnel.

There is a strong, but non-linear relationship between the traffic flow and the accident risk: higher volumes show a higher accident risk (with the exception of very low or very high traffic flows). The extra risk can be counteracted, at least in part, by increasing the light level. This relationship is established for many types of open roads, and it is assumed that it also holds for tunnels.

One of the most important factors is speed. In practice, road and tunnel designs are such that speed and flow usually are interrelated, as a high design speed is selected for roads for which a high flow is expected. High speeds require better visibility and therefore generally a higher luminance level.

The stopping distance is considered with an approximating formula:

$$SD = x_0 + x$$

SD: Stopping distance (m)
x_0 : Reaction way (m)
x : Deceleration way (m)

Reaction way x_0 results as:

$$x_0 = (v/3.6) \cdot t_0$$

v : Speed limit (km/h)
t_0: Reaction time (in design a value of 1.2 is taken into consideration)

The stopping distance results as:

$$x = (v/3.6)^2 / [2g(f \pm s/100)]$$

g: Gravity acceleration $= 9.81$ m/s^2
f: Friction coefficient tire-pavement (in design a value of 0.65 is taken into consideration)
s: Gradient: slope (-) or rise (+) (%)

Speed limit (km/h)	Stopping distance SD at $s = 0$ (m)
100	94
80	65
70	53
60	42
50	32

Table 1: Stopping distance depending on speed limit

4.4   Maintenance factor

In order to maintain the design performance of the lighting installation cleaning is particularly important and walls and luminaires should be washed frequently. The actual cleaning cycle should be related to the maintenance factor used in the calculations of the lighting levels.

The maintenance factor refers to the depreciation in the photometric performance of a luminaire from its state when new to its worst acceptable state in service. At the design stage, a factor of 0.7 is recommended for the maintenance factor in calculation luminance (and illuminances) on the road. This factor applies to the luminaire only.

Lamp output depreciation is accounted for by use of the lamp lumen maintenance factor in the calculation. The initial values should be related to 100 h of use. In practice the luminance of the walls will decrease faster than that of the road surface because both the luminaire output and the reflection factor of the walls will decrease. This can be allowed for at the design stage by assigning a maintenance factor to the wall reflectance, approximately equal to that applied to the luminaire output. Thus the effective maintenance factor used in the calculation of wall luminance is the square of the recommended value of 0.7 i.e. about 0.5. This means that the road luminance in the tunnel should not drop below 0.7 of its initial value and the cleaning cycle should be arranged to ensure that this is the case. The relamping cycle should be monitored in order that the maintenance factor does not fall below 0.7, or that failed lamps give unacceptable uniformity.

4.5   Day-time lighting for long tunnels

4.5.1   Determination of the luminance in the access zone

The visual performance depends on the state of adaptation. L20 can be used for most practical situation as a good approximation of the state of adaptation of the visual system of drivers who approach the tunnel.

Determination of L_{20} using 20° cone

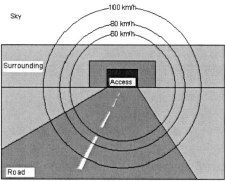

Figure 4: L_{20}-evaluation field

If a 3 dimensional portal view is available, the evaluation of L_{20} is obtained from a sketch of the tunnel entrance surroundings. It is calculated using the formula:

$$L_{20} = \gamma L_C + \rho L_R + \varepsilon L_E + \tau L_{\text{th}}$$

L_C: Sky luminance
γ : Percentage of sky
L_R: Road luminance
ρ : Percentage of road
L_E: Surrounding luminance
ε : Percentage of surrounding
L_{th}: Luminance of threshold zone
τ : Percentage of threshold zone

Traffic direction		North	East-West West-East	South
L_C	Sky	8	12	16
L_R	Road with bitumen surface	3	4	5
	Road with concrete surface	4	5	6
L_E	Rock	3	2	1
	Buildings	8	6	4
	Snow	15 v	10 v	5 v
		15 h	15 h	15 h
	Grass	2	2	2

Table 2: Luminance values [kilo cd/m^2] for different parts of L_{20}-area

(v: vertical area, h: horizontal area)

To estimate the percentage values of the area contributing to the L_{20} value at any tunnel entrance, a photograph is taken from the stopping point. From a known dimension of the picture, e.g. tunnel height, the diameter of the L_{20} cone are determined.

The observer is situated in the centre of right lane, 1.5 m above road surface.

The L_{20} value derived from Table 2 is a maximum value and can be over-estimated since in many tunnel portals the maximum values for the three components from road, sky and surroundings do not occur at the same time at day and year, respectively.

4.5.2   Calculation of L_{20}

L_{20} is calculated using the formula:

$$L_{20} = 2000 \ (\text{cd/m}^2) \cdot k_{\text{sky}} \cdot k_{\text{dd}} \cdot k_{\text{sl}} \cdot k_{\text{ct}}$$

k_{sky}: Factor visible sky (see Table 3)
k_{dd} : Factor driving direction (see Table 4)
k_{sl} : Factor speed limit (see Table 5)
k_{ct} : Configuration of traffic (see Table 6)

Percentage of sky	k_{sky}
0	1.00
10	1.05
25	1.10
35	1.20

Table 3: Factor visible sky

If the percentage of visible sky exceeds 25% the factor k_{dd} is assumed being 2.3.

Driving direction	k_{dd}
North	1.5
East-West	1.7
South	2.0

Table 4: Factor driving direction

Speed limit (km/h)	k_{sl}
up to 70	0.60
above 70	1.00

Table 5: Factor speed limit

Configuration of traffic	k_{ct}
Passenger vehicles and trucks	1.0
Pedestrians, bicycles and motorcycles	1.2

Table 6: Factor configuration of traffic

4.5.3 Measurements of L_{20}

When the lighting equipment of an existing tunnel is to be renewed, L_{20} measurements with installed luminance meters may be undertaken throughout a whole year. L_{20} will be chosen such that this value will be reached or exceeded on 75 hours within one year.

4.6   Lighting in the threshold zone

The necessary lighting level in the threshold zone is determined by visibility criteria being equivalent to enough contrast. A driver shall identify other road users or objects in the threshold zone from the stopping distance.

The total length of the threshold zone is at least equal to the stopping distance. Over the first half of the distance, the luminance level is equal to Lth (the value at the beginning of the threshold zone). Usually from half the stopping distance onwards, the lighting level may gradually and linearly decrease (linear scale) to a value, at the end of the threshold zone, equal to 0.4 L_{th} (see Fig. 5). The gradual reduction over the last half of the threshold zone may also be in steps. However, the luminance levels should not fall below the values corresponding to a gradual decrease, as drawn on the figure.

The average luminance in the first part of the threshold zone (L_{th}) is calculated as follows:

$$L_{th} = k \cdot L_{20} \qquad (L_{th} \text{ and } L_{20} \text{ cd/m}^2)$$

Speed limit (km/h)	k_{CBL}	k_{SYM}
≤ 60	0.035	0.05
80	0.040	0.06
100	0.050	0.10

Table 7: Speed limit k-factor

k_{CBL}: Factor for counter beam lighting
k_{SYM}: Factor for symmetric lighting

4.7   Luminance in the transition zone

The reduction of the luminance of the road in the transition zone follows the curve shown in Fig. 5 resp. the formula:

$$L_{tr} \geq L_{th} \cdot (1.9 + t)^{-1.423}$$

The transition zone starts at the end of the threshold zone $(t = 0)$.

This curve can be replaced by a stepped curve with levels that should never fall below the continuous curve. The maximum luminance ratio permitted on passing from one step to another is 3. The last step should not be greater than 3 times the interior zone luminance.

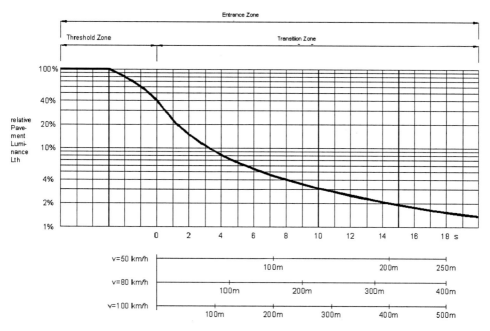

Figure 5: Luminance evolution along the tunnel

4.8 Daytime luminance in the interior zone

The average luminance of the road in the interior zone of the tunnel is given as a function of danger class, speed limit and configuration of traffic.

Tunnels longer than 2.5 km (one-way traffic) or 5 km (two-way traffic) may be divided in two different sub zones. The first sub zone (2.5 km/5 km) corresponds to the formula, the residual sub zone may be operated under normal conditions with 50% of the first sub zone.

$$L_{in} = 6 \ (\text{cd/m}^2) \ \cdot k_{dc} \cdot k_{sl} \cdot k_{td}$$

k_{dc}: Factor of danger class (see table 8)
k_{sl} : Factor speed limit (see table 5)
k_{td}: Factor configuration of traffic (see table 6)

Danger class	k_{dc}
I and II	0.55
III	0.80
IV	1.00

Table 8: Factor danger class

4.9   Luminance in the exit zone

In order to ensure adequate direct illumination of small vehicles and sufficient rear vision via mirrors, the exit zone are illuminated in the same way as the interior zone of the tunnel.

4.10   Parting zone lighting

In case the tunnel is part of an unlit road and the speed of driving is higher than 50 km/h, night-time lighting of the parting zone is recommended:

- If the night-time lighting level in the tunnel is more than 1 cd/m^2;

- If different weather conditions are likely to appear at the entrance and at the exit of the tunnel.

Road lighting in the parting zone shall be provided over the length of two stopping distances with road luminance not lower than 1/3 of the night-time luminance in the interior zone of the tunnel.

4.11   Lighting of the walls

Tunnel walls form part of the background for the detection of obstacles in the tunnel. They contribute to the adaptation level and to the visual guidance. Therefore, the luminance of the tunnel walls is an important component for the quality of the tunnel lighting. The average luminance of the tunnel walls up to at least a height of 2 m, is at least 50% of the average road surface luminance at the relevant location.

4.12   Lighting of lay-by niches

The lighting of lay-by niches is provided according to Figure 6.
1 = luminaires equipped with one lamp NAH 150 W
H2 = luminaires equipped with one lamp HMI 250 W.

All luminaires marked with * are also switched on during night-time and during small traffic.

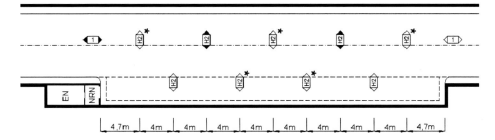

Figure 6: Luminaires at lay-by niches

4.13   Night-time lighting

If the tunnel is on a section of an illuminated road, the quality of the lighting inside the tunnel is at least equal to the level, uniformities and glare of the access road. The uniformity at night of tunnels fulfills the same requirements as the day-time lighting.

If the tunnel is a part of an unlit road, the average road surface luminance inside must not be less than 50% of L_{in}. If the traffic density is lower than 100 vehicles per hour the interior zone may be switched to 25% of L_{in}.

4.14   Emergency lighting

In the event of a failure in the normal power source that supplies the lighting system, an emergency non-interruptible power supply is employed to energize sufficient system luminaires. Conventionally the emergency luminaires form a part of the interior lighting. Emergency lighting provides 25% of interior lighting luminance. The distance between two emergency luminaires must not exceed 40 m.

4.15   Uniformity of luminance

Good uniformity of luminance is provided on the road surface and on the walls up to a height of 2 m. The lower parts of the walls act as a background for traffic, as does the road. So both must be considered in the same way. A ratio of 0.4 for the minimum to the average value of luminance on the road surface and on the walls up to 2 m in height in clean conditions of the tunnel is recommended. A longitudinal uniformity of 0.6 along the center of each lane is recommended for the road. Such values of uniformity must be verified for all dimming stages of the lighting installation. Moreover, in the transition zone, as well as in the second half of the threshold zone (and in the exit zone if existing), the luminance uniformity shall be calculated and measured in the central part of each step replacing the continuous variation curve. It is recommended that the above values shall be reached, independently, on the length of the step.

4.16   Glare restriction

As glare reduces visibility, it is important to minimize it. In tunnel lighting the physiological (disability) glare has to be considered. Disability glare effects are quantified by the Threshold Increment TI as described in CIE 31-1976 "Glare and uniformity in street lighting".

The threshold increment TI must be less than 15% for the threshold, the transition and the interior zones of the tunnel at day-time and night-time. For the exit zone during day-time no restriction is given. The following formula shall be used to calculate TI:

$$TI = 65(L_v/L_{av}^{0.8}) \quad \text{for} \quad L_{av} < 5 \text{ cd/m}^2$$
$$TI = 95(L_v/L_{av}^{1.05}) \quad \text{for} \quad L_{av} > 5 \text{ cd/m}^2$$

L_{av}: Average road surface luminance
L_v : Veiling luminance

The veiling luminance created by all luminaires in the field of view where the axis of fixation is 1° down from the horizontal at the relevant location. The calculations shall be made on the base of the initial values and with a full cut-off angle of 20° above the axis of observation due to the roof of the car. At present, it is not possible to give a numerical value for the restriction of the glare in the transition zone.

4.17   Restriction of the flicker effect

The degree of visual discomfort experienced due to flicker depends upon:

(a) The number of luminance changes per second (flicker frequency);

(b) The total duration of the experience;

(c) The ratio of peak (light) to trough (dark) luminance within each period (luminance modulation depth) and the steepness in the increase (rise-time).

(a), (b) and (c) depend upon vehicle speed and luminaire spacing. (c) also depends upon the photometric characteristics and the spacing of the luminaires. In nearcontinuous line lighting, where the distance between the end of one luminaire and the beginning of the next luminaire is less than the length of the luminaires, flicker discomfort is rendered independent of the frequency. The flicker frequency can be easily established by dividing the velocity (in m/s) by the luminaire spacing (center-to-center; in m). For example: for a speed of 60 km/h (= 16.6 m/s) and luminaire spacing of 4 m, the flicker frequency is 16.6/4 = 4.2 Hz.

In general, the flicker effect is negligible at frequencies below 2.5 Hz and above 15 Hz. When the frequency is between 4 Hz and 11 Hz, and has duration of more than 20 s, discomfort may arise provided no other measures are taken. It is recommended that, in installations where the duration is more than 20 s, the frequency range between 4 Hz and 11 Hz be avoided, particularly when small light sources with a sharp run-back are used. Large size luminaires with low gradients in the light distribution (like e.g. length-wise mounted luminaires with fluorescent tubes) usually will lead to little discomfort. In view of the high luminance of the elements, it is recommended to avoid for non-suntight daylight screens all frequencies below the flicker-fusion frequency, e.g. > 50 Hz, independent of the length of the screens.

$$FF = v/3.6 \cdot y$$

v: Speed limit (km/h)
y: Luminaire distance (m)

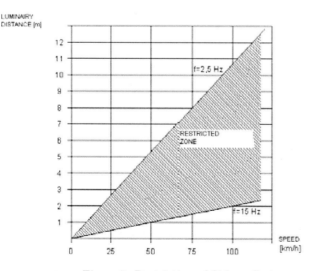

Figure 7: Restriction of flicker effect

4.18   Daylight variation and lighting control

The access zone luminance varies with changes in daylight conditions.

As the luminance levels in the threshold and transition zones are constant percentages of the access zone luminance, it is necessary to provide control of the lighting in these zones. When screened daylight is used for the entrance zone lighting, the control is automatic, It should be noted, however, that, depending on the transmission characteristics of the screens, the luminance under the screens, is not always linearly correlated to the outdoor light level. For artificial lighting a system that provides control is needed. The control may be done through continuously dimming devices or by switching in 8 steps.

For adequate light control the access zone light level must be monitored continuously.

The luminance meter that is applied for these measurements is placed at the stopping distance and aimed at the tunnel portal. For maintenance reasons, the luminance meter should be mounted between 2 and 5 m high above the pavement or hard shoulder on the near side of the street if the road does not curve towards the near side. In the latter case, the luminance meter should be positioned over the central reservation or on the off side of the road. Its sensitivity to temperature should be minimal and the long-term stability should be good as well.

From the momentary access luminance value, the instantaneous luminance required in the threshold zone (L_{th}) is derived. In order to be able to correct for short-term disturbances (e.g. the burning out of lamps) as well as the long-term deterioration (e.g. soiling and corrosion of the tunnel walls and the luminaires) it is recommended to measure the interior luminance with a (second) luminance meter. Again here, it is not possible to measure the luminance from the position of the driver s eye. For practical reasons the mounting height of the luminance meter shall be 3.0 m above surface. Therefore the measured value is different from the luminance seen by the driver. A (constant) correction factor has to be applied.

In general, the difference between the actual and the preferred luminance in the threshold zone should be as small as possible. For each installation, the most economic solution has to be considered on the basis of the energy, lamp and labour cost.

5 Measurement of tunnel lighting installation

5.1 Quality numbers for tunnel lighting installations

To describe quantitatively the performance of a tunnel lighting installation, the following parameters should be evaluated:

- Average road luminance in all zones

- Average luminance of walls in all zones

- Average road illuminance in all zones

- Overall uniformity on road

- Overall uniformity on walls

- Longitudinal uniformity on road

- Longitudinal uniformity on wall

- Threshold increment

These parameters should be measured with highest stage (100%) as well as 50% stage.

5.2 Measuring fields

For comparison of calculation and field measurements the calculation should have the same grid as the measurement grid.

Uniformity values are strongly depended on the position and the number of the grid point.

For tunnel sections with constant lighting levels (first part of entrance zone and interior zone) the measuring grid shall be as shown in Fig. 8. In zones with variable lighting levels the measuring grid shall be in longitudinal distances of 6 m with three measuring grids per lane.

The length of the measuring field shall be between two or more luminaires and shall have a minimum length of 20 m.

The observer is 60 m in front of the measuring field in the centre of the right lane and in a height of 1.5 m above surface.

The luminance meter shall have an acceptance angle of 2' vertical and 20' horizontal, for measuring the wall luminance an acceptance angle of 20° is sufficient.

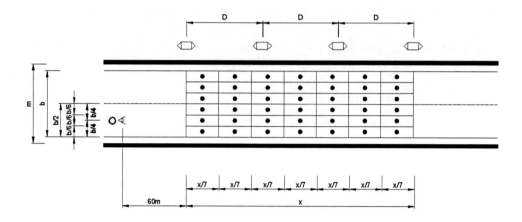

Figure 8: Calculation and measuring grid of road surface

O: Observer
D: Luminaire spacing
x : Length of measuring field
b : Width of lane
m: Distance between walls

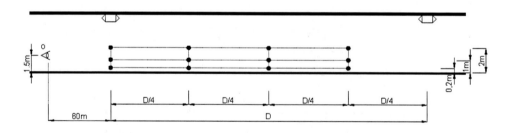

Figure 9: Calculation and measuring grid of wall

References

[1] RVS 9.27 – Projektierungsrichtlinien Beleuchtung

[2] CIE 88:2004, 2nd edition: International Commission on Illumination guide for the lighting of road tunnels and underpasses

[3] CIE 31-1976 – Glare and uniformity in street lighting

[4] CEN Report CR 14380, April 2003 - Lighting applications – Tunnel lighting

Rock reinforcing

Dimitrios Kolymbas

Institute of Geotechnical and Tunnel Engineering, University of Innsbruck, Technikerstr. 13, A-6020 Innsbruck, Austria

Abstract: As a contrast to the empirical application of rockbolts in tunnelling, a rational approach is attempted here. Pattern bolting is modelled under the assumption of homogeneously distributed bolts. Pre-stressed and non-prestressed bolts are considered and simple equations are derived that enable to quantify their contribution to the tunnel lining.

1   Introduction

Anchors or rockbolts are reinforcements (usually made of steel) which are inserted into the ground to increase its stiffness and strength. There are various sorts of reinforcement actions and the corresponding terminology is not uniform (see also [1]). The following terminology is used in soil mechanics: If the reinforcement bar is fixed only at its both ends, then it is called an 'anchor'. Anchors can be pre-stressed or not, in the latter case they assume force only after some extension (e.g. due to convergence of the tunnel). If the reinforcement bar is connected to the surrounding ground over its entire length, then it is called a 'nail' or 'bolt'. The connection can be achieved with cement mortar. Following the general use in tunnelling, in this article the words 'anchor', 'nail' and 'rockbolt' are considered as synonyms. In jointed rock, reinforcement bars are placed ad hoc to prevent collapse of individual blocks. Anchoring or bolting in a regular array is called 'pattern bolting'.

2   Impact of pattern bolting

It is generally believed that reinforcing improves the mechanical behaviour of ground. Despite several attempts however, the reinforcing action of stiff inlets is not yet satisfactorily understood and their application is still mainly empirical. In civil engineering, reinforcement has been first applied in reinforced concrete, where it takes over the tensile stress. In geotechnical engineering, however, tensile stresses are in general not present and the mere existence of stiff inlets does not necessarily increase the stiffness and strength of the compound material. Of paramount importance is the force transmittance

from the surrounding ground to the reinforcement. In some approaches, reinforced ground is considered as a two-phase continuum in the sense that both
constituents are assumed to be smeared and present everywhere in the considered body. Thus, their mechanical properties prevail everywhere, provided
that they are appropriately weighed (Section 4).

The stiffening action of inlets can be demonstrated if we consider a conventional triaxial test on a soil sample containing a thin pin of, say, steel (Fig. 1).
The stiff inlet is here assumed as non-extendable (i.e. rigid). Therefore, its
vertical displacement is constant as shown in Fig. 1 left. This implies a relative slip of the adjacent soil, which is oriented downwards in the upper half
and upwards in the lower half. Being stiffer, the pin 'attracts' force and, thus,
the adjacent soil is partly relieved from compressive stresses. As a result, the
triaxial sample, viewed as a whole, is now stiffer. This effect is closely related
to 'tension stiffening' known in concrete engineering.

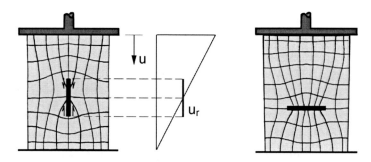

Figure 1: Steel inlet in triaxial sample, distribution of vertical displacements and stress
trajectories for two different orientations of the inlet.

Obviously, the length of the stiff inlet plays an important role, since the
transmission of shear forces between ground and inlet can only be mobilized
by a relative slip which imposes to the compound material an interal length.
This fact is particularly well illustrated in the case of steel fibres reinforced
shotcrete. There, the steel fibres do not substantially affect the behaviour of
shotcrete unless cracks appear.

Another way to increase the stiffness of reinforced soil is given by pre-stressing,
i.e. increasing the pressure level. As known, the stiffness of granular materials
increases almost linearly with stress level. The latter can be increased by pre-
stressing an array of anchors, i.e. of reinforcing inlets that transmit the force
to the surrounding ground only at their ends ad not over their entire length.
An analysis of this mechanism is presented in the next section.

3 Ground stiffening by pre-stressed anchors

The strengthening effect of pre-stressed pattern bolting will be considered for the case of a tunnel with circular cross section within a hydrostatically stressed elastoplastic ground. The primary hydrostatic stress is σ_∞. If the spacing of the anchors is sufficiently small, their action upon the ground can be approximated with a uniform radial stress σ_A (Fig. 2).

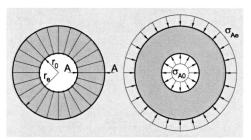

Figure 2: Idealised pattern bolting

The radial stress σ_A is obtained by dividing the anchor force with the pertaining surface. Let n be the number of anchors per one meter of tunnel length. We then obtain

$$\sigma_{A0} = \frac{nA}{2\pi r_0} \quad , \quad \sigma_{Ae} = \frac{nA}{2\pi r_e}$$

or

$$\sigma_{Ae} = \sigma_{A0} \cdot \frac{r_0}{r_e} \quad .$$

It is, thus, reasonable to assume the following distribution of σ_A within the range $r_0 < r < r_e$

$$\sigma_A = \sigma_{A0} \cdot \frac{r_0}{r} \quad . \tag{1}$$

We consider the entire stress in the range $r_0 < r < r_e$. Pre-stressing of the anchors increases the radial stress from σ_r to $\sigma_r + \sigma_A$.

We now assume that in the range $r_0 < r < r_e$ the shear strength of the ground is fully mobilised. For this case we will determine the support pressure p. For simplicity, we consider a cohesionless ground ($c = 0$) and obtain

$$\sigma_\theta = K_p(\sigma_r + \sigma_A) \tag{2}$$

with $K_p = \dfrac{1 + \sin\varphi}{1 - \sin\varphi}$. Equilibrium in radial direction reads

$$\frac{d(\sigma_r + \sigma_A)}{dr} + \frac{\sigma_r + \sigma_A - K_p(\sigma_r + \sigma_A)}{r} = 0 \quad . \tag{3}$$

Introducing (1) into (3) yields

$$\frac{d\sigma_r}{dr} + \frac{1}{r}\left[\sigma_r(1 - K_p) - K_p\sigma_{A0}\frac{r_0}{r}\right] = 0 \quad . \tag{4}$$

The solution of the differential equ. (4) is obtained as $\sigma_r = \text{const} \cdot r^{K_p-1} - \sigma_{A0}\frac{r_0}{r}$. The integration constant is obtained from the boundary condition $\sigma_r(r_0) \stackrel{!}{=} p$ where p is the pressure exerted by the ground upon the lining. We finally obtain

$$\sigma_r = (p + \sigma_{A0}) \cdot \left(\frac{r}{r_0}\right)^{K_p-1} - \sigma_{A0}\frac{r_0}{r} \quad . \tag{5}$$

At the boundary of the elastic region (at $r = r_e$) it must be $\sigma_r = \sigma_e$, where σ_e is obtained from equ. $\sigma_e = 2\sigma_\infty/(K_p + 1)$ [2]:

$$(p + \sigma_{A0})\left(\frac{r_e}{r_0}\right)^{K_p-1} - \sigma_{A0} \cdot \frac{r_0}{r_e} = \frac{2}{K_p + 1} \cdot \sigma_\infty \tag{6}$$

We meet the simplifying assumption that the plastified zone coincides with the anchored ring, i.e. we introduce $r_e = r_0 + l$, where l is the theoretical anchor length, into equ. (6) and eliminate p. We thus obtain the support pressure in dependence of the pre-stressing force A of the anchors, their number n per tunnel meter, the theoretical anchor length l, the tunnel radius r_0, the primary stress σ_∞ and the friction angle φ:

$$p = \left(\frac{2\sigma_\infty}{K_p + 1} + \frac{nA}{2\pi r_0} \cdot \frac{r_0}{r_0 + l}\right)\left(\frac{r_0}{r_0 + l}\right)^{K_p-1} - \frac{nA}{2\pi r_0} \tag{7}$$

The real anchor length L should be greater than the theoretical one, in such a way that the anchor force can be distributed along the boundary $r = r_e$ (Fig. 3). In practice, the anchor lengths are taken as 1.5 to 2 times the thickness of the plastified zone.

To consider cohesion, equ. (2) is replaced by

$$\sigma_\theta = K_p(\sigma_r + \sigma_A) + 2c\frac{\cos\varphi}{1 - \sin\varphi} \quad .$$

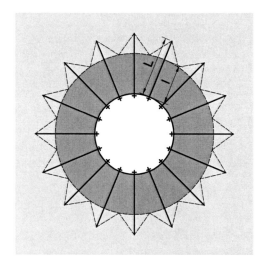

Figure 3: Theoretical (l) and real (L) anchor lengths

Thus, equilibrium in radial direction reads

$$\frac{d\sigma_r}{dr} + \frac{1}{r} \cdot \left[\sigma_r(1 - K_p) - 2c\frac{\cos\varphi}{1 - \sin\varphi} - K_p\sigma_{A0}\frac{r_0}{r} \right] = 0 \quad .$$

The solution of this differential equation reads:

$$\sigma_r = \text{const} \cdot r^{K_p - 1} - \sigma_{A0}\left(\frac{r_0}{r}\right) - c\frac{2\cos\varphi}{(K_p - 1)(1 - \sin\varphi)} \quad .$$

With the boundary condition $\sigma_r(r_0) \overset{!}{=} p$ and with $\frac{2\cos\varphi}{(K_p-1)(1-\sin\varphi)} = \cot\varphi$ one finally obtains

$$\sigma_r = (p + \sigma_{A0} + c \cdot \cot\varphi)\left(\frac{r}{r_0}\right)^{K_p - 1} + \sigma_{A0}\frac{r_0}{r} - c \cdot \cot\varphi \quad .$$

From the requirement $\sigma_r(r_e) = \sigma_e$ with σ_e according to

$$\sigma_e = \sigma_\infty(1 - \sin\varphi) - c\cos\varphi \quad [2]$$

it is obtained:

$$(p + \sigma_{A0} + c \cdot \cot\varphi)\left(\frac{r_e}{r_0}\right)^{K_p - 1} - \sigma_{A0}\frac{r_0}{r_e} - c \cdot \cot\varphi$$
$$= \sigma_\infty(1 - \sin\varphi) - c \cdot \cos\varphi \quad ,$$

With $r_e = r_0 + l$ it finally follows:

$$p = \sigma_\infty(1 - \sin\varphi)\left(\frac{r_0}{r_0 + l}\right)^{K_p-1} - \frac{nA}{2\pi r_0}\left[1 - \left(\frac{r_0}{r_0 + l}\right)^{K_p}\right]$$
$$-c \cdot \cot\varphi\left[1 - \left(\frac{r_0}{r_0 + l}\right)^{K_p-1}\right] - c \cdot \cos\varphi\left(\frac{r_0}{r_0 + l}\right)^{K_p-1} \quad . \tag{8}$$

If the ground pressure is to be taken solely by the anchors (i.e. $p = 0$), then:

$$nA \geq \frac{2\pi r_0}{1 - \left(\frac{r_0}{r_0 + l}\right)^{K_p}} \cdot \left\{\sigma_\infty(1 - \sin\varphi)\left(\frac{r_0}{r_0 + l}\right)^{K_p-1}\right.$$
$$\left.-c \cdot \cot\varphi\left[1 - \left(\frac{r_0}{r_0 + l}\right)^{K_p-1}\right] - c \cdot \cos\varphi\left(\frac{r_0}{r_0 + l}\right)^{K_p-1}\right\} \quad . \tag{9}$$

A disadvantage of equations (8), (9) and all other approaches where the action of anchors is smeared (homogenized) is that the number n and the force A appear only in the product nA. Thus, it cannot be decided how many anchors are really needed.

In tunnelling, anchors are usually not pre-stressed. Thus the analysis presented in this section is rather academic.

4   Transmission of shear stress

In this section we consider the stiffening effect of arrays of bolts, i.e. reinforcing elements that are not pre-stressed and transmit shear forces to the surrounding ground over their entire length. Considering equilibrium of the normal force N and the shear stress τ applying upon the periphery of a bolt element of the length dx (Fig. 4) we obtain $dN = \tau\pi d \cdot dx$. With $N = \sigma\pi d^2/4$, $\sigma = E\varepsilon$ and $\varepsilon = du_s/dx$ we obtain

$$\frac{d^2 u_s}{dx^2} = \frac{4\tau}{Ed} \quad , \tag{10}$$

with u_s being the displacement of the bolt. Obviously, the shear stress τ acting between bolt and surrounding ground is mobilised with the relative displacement, $\tau = \tau(s)$, $s = u_s - u$, where u is the displacement of the ground.[1]

---

[1] Consider e.g. the relations used in concrete engineering [3].

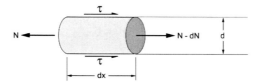

Figure 4: Forces upon a bolt element

Of course, u depends on τ: In a first step of simplified (uncoupled) analysis we assume that u does not depend on τ and is given by the elastic solution for the convergence of a circular tunnel in elastic ground with constant hydrostatic primary stress (see e.g. [2]):

$$u = \frac{\sigma_\infty - p}{2G} \frac{r_0{}^2}{r} \quad .$$

Herein, r is the radius with respect to the tunnel axis. Furthermore, we assume a rigid-idealplastic relation $\tau(s)$, i.e. τ achieves immediately its maximum value τ_0. Thus, the total force transmitted by shear upon a bolt of the length l is $l\tau_0\pi d$. This force is applied via the top platen upon the tunnel wall. Assuming n bolts per m^2 tunnel wall we obtain thus the equivalent support pressure $p_{bolt} = nl\tau_0\pi d$. If the arrangement of bolts is given by the spacings a and b (Fig. 5), then $n = 1/(ab)$. Thus,

$$p_{bolt} = \frac{1}{ab}\tau_0\pi d l \tag{11}$$

modifies the support line as shown in Fig. 6. It is remarkable to notice that τ_0 is limited by the shear strength τ_R of the rock, $\tau_0 \leq \tau_R$.[2] Therefore, reinforcement cannot considerably improve the stiffness/strength of very weak rock. Equally, reinforcement is ineffective if it is placed in the direction of the lines of zero-extension.

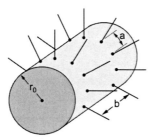

Figure 5: Array of bolts

---

[2]E.g., for marl and hard clay $\tau_o \approx (0.45\text{-}0.60)c_u$ (cited in [4])

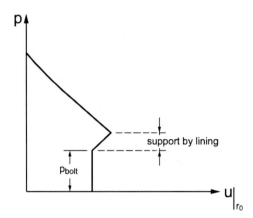

Figure 6: Ground reaction line and support line affected by idealised bolts (Assumptions: rigid bolts, rigid-idealplastic shear stress transmission to the ground, ground displacement not influenced by the bolts, installation of bolts is instantaneous).

5   Multiphase model of reinforced ground

For multiphase media (such as composite materials) it is assumed that each point is occupied by all constituents, which in our case are ground and reinforcement. The corresponding field quantities are denoted by the indices g and r, respectively. Let us consider a representative volume element (REV) with the volume V. The included volumes of ground and reinforcement are V_g and V_r, respectively. The corresponding volume fractions are $\alpha_g := V_g/V$ and $\alpha_r = V_r/V$ with $\alpha_g + \alpha_r = 1$. It can be shown that the volume fractions are equal to the area fractions, e.g. $\alpha_g = A_g/A$ with A_g being the cross section occupied by ground, and A being the total cross section. Referring to multiphase media and quantities (such as density and stress) of the several constituents, it has to be distinguished between the 'real' (or 'effective') quantities, that prevail (or are averaged) over the individual phases and the 'partial' quantities, that are averaged over the entire REV. Thus, we have the real densities of ground and reinforcement, ρ^g and ρ^r, respectively. The corresponding partial densities are $\rho_g = \alpha_g \rho^g$ and $\rho_r = \alpha_r \rho^r$. Similarly, it has to be distinguished between the stresses σ^g and σ^r on the one hand, and $\sigma_g = \alpha_g \sigma^g$ and $\sigma_r = \alpha_r \sigma^r$ on the other.[3] In the quasi-static case (i.e. accelerations are negligible) the equations of equilibrium for the two phases read:

$$\nabla \cdot \sigma_g + \mathbf{P}_{rg} + \rho_g \mathbf{g} = 0$$
$$\nabla \cdot \sigma_r + \mathbf{P}_{gr} + \rho_r \mathbf{g} = 0 \qquad (12)$$

---

[3]In the notation of multiphase media, σ^g is the 'effective' stress in the ground. This quantity should *not* be confused with the effective stress in the usual sense of soil mechanics.

g is the gravity acceleration, $\mathbf{P}_{rg} = -\mathbf{P}_{gr}$ is a vector that characterises the interaction of the two phases. $\mathbf{P}_{rg}$ is the force per unit volume exerted by the reinforcement upon the ground. For the case of the fully mobilised rigid-idealplastic shear stress τ_0 the interaction force can be determined as follows: The shadowed volume (Fig. 7)

$$V_0 = \int_0^{\theta_0} \int_0^b \int_{r_0}^{r_0+l} r\,dr\,dz\,d\theta \quad,$$

with $\theta_0 = a/r_0$, corresponds to one bolt. Thus, the interaction force $\pi d\tau_0 l$ is obtained as:

$$\pi dl\tau_0 = \int_0^{\theta_0} \int_0^b \int_{r_0}^{r_0+l} P_{rg}r\,dr\,dz\,d\theta \quad. \tag{13}$$

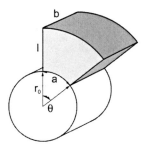

Figure 7: Volume corresponding to one bolt

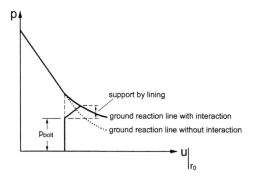

Figure 8: Ground reaction line altered by the action of bolts and corresponding support line.

Knowing that P_{rg} depends on r, we set $P_{rg} = \text{const}/r$ and obtain from equ. (13):

$$P_{rg} = \frac{\pi d\tau_0 r_0}{ab} \cdot \frac{1}{r} \quad.$$

Obviously, for very thin bolts ($d \rightarrow 0$) the interaction force P_{rg} becomes negligible and, thus, its neglection in Sec. 4 becomes justified. A non-vanishing interaction force can be taken into account as follows: As usual for deep tunnels, we neglect gravity ($\mathbf{g} \approx \mathbf{0}$). Then, equ. (12) written in cylindrical coordinates reads:

$$\frac{d\sigma_r}{dr} + \frac{\sigma_r - \sigma_\theta}{\theta} = -\frac{\pi d \tau_0 r_0}{ab} \cdot \frac{1}{r} \tag{14}$$

Introducing the MOHR-COULOMB yield condition leads to

$$\frac{d\sigma_r}{dr} - \frac{(\sigma_\theta + \sigma_r)\sin\varphi}{r} - \frac{2c \cot\varphi}{r} = -\frac{\pi d \tau_0 r_0}{ab} \cdot \frac{1}{r} \quad . \tag{15}$$

We can reduce equ. (15) to the original equation, that holds for ground without anchors, if we replace the cohesion c by $\hat{c}$, where

$$\hat{c} := c - \tau_0 \frac{\pi d r_0}{2ab} \cdot \tan\varphi. \tag{16}$$

Thus, the action of the bolts upon the ground is equivalent to a reduction of cohesion according to equ. (16). Note that equ. (16) holds for $\varphi > 0$. For $\varphi = 0$ and $\sigma_\theta - \sigma_r = 2c$ the equilibrium equ. (15) has to be replaced by

$$\frac{d\sigma_r}{dr} - \frac{2c}{r} = -\frac{\pi d \tau_0 r_0}{ab} \cdot \frac{1}{r} \quad .$$

The solution (14.30) will be valid if we replace c by $\hat{c}$, where

$$\hat{c} := c - \tau_0 \frac{\pi d r_0}{2ab} \quad .$$

A perusal of equ. (12) shows that the action of bolts upon the ground is equivalent to the action of an additional volume force which is directed towards the centerline of the tunnel. The volume force due to the action of the bolts alters the ground reaction line as shown in (Fig. 8). The resulting ground reaction line implies higher pressure upon the lining. In this sense the action of bolts is adverse. This effect, however, is more than compensated by the additional support p_{bolt}.

Some additional specification must be given to apply the equations given so far:

Diameter d of reinforcement: For grouted rockbolts ('SN-anchors') it should be distinguished between the diameter of the steel bar (d_1) and the one of the grouted borehole (d_2). If τ_0 is the maximum shear stress transmittable from the grouted borehole to the adjacent ground, then $d = d_2$ is to be set in equ. (16). To check the tensile stress in the steel rod, the derivation of equ. (10) must be re-considered. Assuming that the hardened mortar cannot transfer any tensile stresses (it may, thus, be fissured), we have $dN = \tau \pi d_2 \cdot dx$ and $N = \sigma \pi d_1^2/4$ leading to $d^2 u_s/dx^2 = d_2/d_1^2 \cdot 4\tau/E$.

Length l of rockbolts: If l is a-priori prescribed then the derivation of the analytical expression for the ground reaction is complicated. It is much simpler to assume that l equals the width of the plastified zone around the tunnel and to use the known expression (see e.g. [2]) for the convergence $u|_{r_0}$:

Example

Tunnel: radius $r = 6$ m

Rock:[4] Young's modulus $\quad E = 2$ GPa $\rightsquigarrow G = \dfrac{E}{2(1+\nu)} = \dfrac{E}{3} = 667$ MPa

$$\begin{aligned} \text{friction angle} \quad & \varphi = 30° \\ \text{dilatancy} \quad & b = 0 \\ \text{cohesion} \quad & c = 250 \text{ kPa} \\ \text{primary stress} \quad & \sigma_\infty = 18 \text{ MPa} \end{aligned}$$

Shotcrete:[5] Young's modulus $\quad E = 11100$ N/mm$^2 = 11$ GPa (8 days)

$\qquad\qquad$ uniaxial strength $\quad q_u = 28$ N/mm$^2 = 28$ MPa (8 days)

$\qquad\qquad$ thickness $\qquad\qquad d = 0.4$ m

Rockbolts:[6] $\quad l = 8$ m

$\qquad\qquad d = 6$ cm

$\qquad\qquad \tau_0 = 0.3$ MN/m^2

$\qquad\qquad a = b = 1.5$ m

A. Situation without rockbolts

$$u|_{r_0} = r_0 \left[\frac{\sigma_\infty(1 - \sin\varphi) - c(\cos\varphi - \cot\varphi)}{p + c\,\cot\varphi} \right]^{\frac{2-b}{(K_p-1)(1-b)}} \cdot \frac{\sigma_\infty}{2G} \left(\sin\varphi + \frac{c}{\sigma_\infty}\cos\varphi \right)$$

---

[4] see e.g. [5]
[5] see e.g. [6]
[6] see e.g. [7]

Ground reaction line:

$$u|_{r_0} = r \underbrace{\left[\frac{18\,(1-0.5) - 0.25\,(0.866 - 1.732)}{p + 0.25 \cdot 1.732}\right]^1}_{9.216} \cdot \underbrace{\frac{18}{2 \cdot 667}\left(0.5 + \frac{0.25}{18} \cdot 0.866\right)}_{0.00691}$$

$$= r \cdot \frac{0.0637}{p\,(\text{MPa}) + 0.433} \quad \text{for } p \le p^*$$

Determination of p^* for rock with $\varphi > 0,\ c > 0$:

$$r_e = r_0 \left(\frac{\sigma_\infty\,(1-\sin\varphi) - c\,(\cos\varphi - \cot\varphi)}{p + c\cot\varphi}\right)^{\frac{1}{K_p-1}} \tag{17}$$

From equ. (17) it follows for $r_e = r_0$:

$$\begin{aligned}
p^* &= \sigma_\infty(1-\sin\varphi) - c\,(\cos\varphi - \cot\varphi) - c\cot\varphi \\
&= \sigma_\infty(1-\sin\varphi) - c \cdot \cos\varphi \\
&= 18 \cdot 0.5 - 0.25 \cdot 0.866 = 8.78 \text{ MPa}
\end{aligned}$$

Ground reaction line for $p > p^*$:

$$u|_{r_0} = r \cdot \frac{\sigma_\infty}{2G}\left(1 - \frac{p}{\sigma_\infty}\right) = r \cdot \frac{18}{2 \cdot 667}\left(1 - \frac{p}{18}\right) = r \cdot 0.0135 \left(1 - \frac{p}{18}\right)$$

Support reaction line:

$$u\,(\text{m}) = u_0 + \frac{r^2}{Ed}\,p = u_0 + \frac{36}{11 \cdot 10^3 \cdot 0.4}\,p = u_o\,(\text{m}) + 8.18 \cdot 10^{-3} p\,(\text{MPa})$$

The strength of shotcrete requires:[7]

$$p \overset{!}{\le} \frac{d}{r}\,q_u = \frac{0.4}{6} \cdot 28 = 1.87 \text{ MPa} \tag{18}$$

Thus, the reaction lines read:

Ground:

$$u|_{r_0}\,(\text{cm}) = \begin{cases} 8.1\left(1 - \dfrac{p}{18}\right) & \text{for } p \ge 8.78 \text{ MPa} \\[2mm] \dfrac{38.21}{p + 0.433} & \text{for } p < 8.78 \text{ MPa} \end{cases}$$

---

[7] For simplicity, safety factors are not considered here.

Support:

$$u|_{r_0}(\mathrm{m}) = u_0(\mathrm{m}) + 8.18 \cdot 10^{-3} p \ (\mathrm{MPa})$$

The inverse relation of the ground line reads:

$$p = \begin{cases} 18 - 2.222\, u|_{r_0} & \text{for} \quad u|_{r_0} \leq 4.149 \ \mathrm{cm} \\ \dfrac{38.21}{u} - 0.433 & \text{for} \quad u|_{r_0} > 4.149 \ \mathrm{cm} \end{cases}$$

The inverse relation of the support line reads: $p = (u - u_0) \cdot 122.$

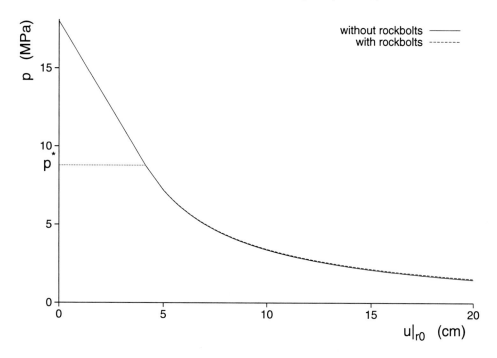

Figure 9: Ground reaction lines without and with rockbolts.

If the strength of shotcrete is to be completely exploited (i.e. safety factor=1):

$$p_{\text{ground}} = \frac{38.21}{u} - 0.433 \stackrel{!}{=} 1.87 \tag{19}$$

$$\rightsquigarrow u = \frac{38.21}{1.87 + 0.433} = 16.59 \ \mathrm{cm}$$

$$= u_0 + 8.18 \cdot 10^{-1} \cdot 1.87 \ \mathrm{cm}$$

$$\rightsquigarrow u_0 = 15.06 \ \mathrm{cm}$$

This means that shotcrete can be only applied after a convergence of 15.06 cm.

B. Situation with rockbolts:

$$p_{\text{bolt}} = \frac{1}{ab}\tau_0 \pi dl = \frac{1}{1.5\text{m} \cdot 1.5\text{m}} \cdot 0.3 \cdot \pi \cdot 0.06 \cdot 8 = 0.200 \text{ MPa}$$

The cohesion of the rock is reduced:

$$\hat{c} = c - \tau_0 \frac{\pi dr}{2ab} \tan \varphi$$
$$= 0.25 - 0.3 \frac{\pi \cdot 0.06 \cdot 6}{2 \cdot 2.25} \tan 30°$$
$$= 0.206 \text{ (MPa)}$$

Herewith:

$$p^* = 18 \cdot 0.5 - 0.206 \cdot 0.866 = 8.82 \text{ MPa}$$

Ground reaction line for $p < p^*$:

$$u|_{r_0} = r \left[\frac{\overbrace{18 \cdot 0.5 - 0.206(0.866 - 1.732)}^{9.178}}{\underbrace{p + 0.206 \cdot 1.732}_{0.357}} \right] \cdot \underbrace{\frac{18}{2 \cdot 667}\left(0.5 + \frac{0.206}{18} \cdot 0.866\right)}_{0.00688}$$

$$= \frac{37.89}{p + 0.357}$$

Inverse relation: $p = \dfrac{37.89}{u} - 0.357$ for $p < 8.82$ MPa

If the strength of shotcrete is to be completely exploited:

$$p - p_{\text{bolt}} = 1.87 \text{ MPa}$$
$$\frac{37.89}{u} - 0.357 - 0.200 = 1.87 \rightsquigarrow u = 15.60 \text{ cm} \qquad (20)$$
$$15.60 = u_0 + 8.18 \cdot 10^{-1} \cdot 1.87$$

$\rightsquigarrow$ Shotcrete can be applied after a convergence of 14.07 cm.
I.e., the improvement due to rockbolts in this case is marginal.

In this example the thickness of the shotcrete was chosen unusually large (40 cm). Let us now consider, how the results change if we choose the more usual thickness of 30 cm:

Equ. (18) then yields a maximum ground pressure of $p = 1.403$ MPa, which results via equ. (19) in $u = \frac{38.21}{1.403 + 0.433} = 20.81$ cm leading thus to a pre-convergence of $u_0 = 19.66$ cm for the case without rockbolts.

For the case *with* rockbolts we obtain from equ. (20): $u = 19.33$ cm and thus the resulting pre-convergence of $u_0 = 18.18$ cm.

6 Reinforcement of the face

Temporal reinforcement of the face is an alternative to partial face excavation. This technique is applied in deep tunnels, whereas in shallow tunnels in soil the face is usually supported by pressurized slurry or earth pressure balance. As in equ. (11), the support pressure exerted on the face by the rockbolts reads $p_{bolt} = n\tau_0\pi d_2 l$, where n is the number of bolts per unit area of the tunnel face. A statically necessary face support pressure is hard to be specified. It appears reasonable that the main task of the face bolts is to render the rock ductile so that the rock at the face can sufficiently yield in vertical direction without caving into the excavated space.

References

[1] C.R. Windsor, A.G. Thompson: Rock Reinforcement - Technology, Testing, Design and Evaluation. In: Comprehensive Rock Engineering, Vol. **4**, Pergamon Press 1993, pp. 451-484

[2] D. Kolymbas, Tunnelling and Tunnel Mechanics, Springer 2005 (in print)

[3] K. Zilch and A. Rogge: Grundlagen der Bemessung von Beton-, Stahlbeton- und Spannbetonbauteilen nach DIN 1045-1. In: Betonkalender 2000, BK1, 171-312, Ernst & Sohn Berlin, 2000

[4] H. Ostermayer: Grundbau Taschenbuch Teil 2, Ernst & Sohn 2001, p. 189

[5] K. Kovari, F. Amberg, H. Ehrbar, Mastering of Squeezing Rock in the Gotthard Base, World Tunnelling 2000

[6] LAICH SA, Spritzbeton, Avegno 1991

[7] H. Ostermayr, Verpressanker, Grundbau Taschenbuch Teil 2, Ernst & Sohn, Berlin, 2001, p.191

Tunnelling with soil freezing

Wolfgang Orth

Dr.-Ing. Orth GmbH, Ingenieurbüro für Bodenmechanik und Grundbau, Tiroler Straße 7, D-76227 Karlsruhe, Germany

Abstract: For the use in tunnelling the basics of soil freezing and the behaviour of frozen soil is briefly described. The application of soil freezing is shown at four executed tunnel projects with different freezing pipe arrangements and shapes of the frozen bodies.

1 The principle of soil freezing

The principle of ground freezing is to make the soil solid and watertight by transforming the pore water into ice. This idea is not new. The first patent in Germany dates back to 1883, just 7 years after the invention of the compressor cooling machine by Carl von Linde. At the same time a lot of freezing activities also started mainly in England and Belgium, in the beginning all of them to sink mining shafts.

2 How to freeze soil

To extract heat out of the soil, freezing pipes are drilled into the soil. Each pipe contains a second pipe inside to feed a cooling fluid, that cools first the exterior pipe and then the surrounding soil.

A frozen column is gradually formed around each pipe until it touches the neighbour frozen columns. If the pipes are drilled vertically in a straight row the frozen columns finally form a vertical plane wall, or a cylinder wall if they are positioned on a circle. For tunnelling, the freezing pipes can be drilled horizontally into the working face and so form a shell of frozen soil.

The cold fluid extracting the heat from the soil can be a brine which is cooled in a compressor cooling machine. Another possibility is to feed the pipes with liquid nitrogen, which is brought to the side in insulated tanks carried by trucks.

The frost propagation around the pipes depends not only on time, but also on the soil type, bulk density, water content and the initial soil temperature.

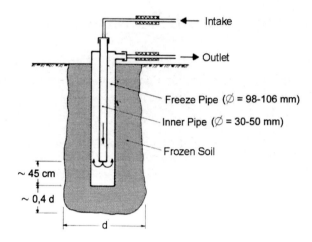

Figure 1: Freezing pipe

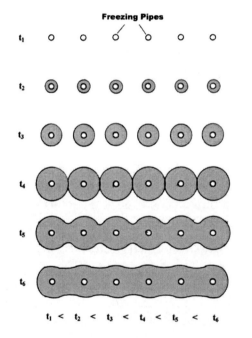

Figure 2: Frost propagation around the freezing pipe

As the so-called latent heat of water, what is the energy that has to be extracted to transform water of 0°C to ice of 0°C, is comparatively high, the water content strongly affects the frost propagation.

Flowing water adds thermal energy to the freezing system and this additional thermal energy must be removed by the freezing pipes. So the presence of flowing water can significantly delay or even prevent the development of ice.

3   The mechanical behaviour of frozen soil

The strength of frozen soil differs from that of rock or concrete as it is a visco-plastic material. Frozen soil shows significant time dependent deformations under constant load, i.e. creeping. The creep rate increases with the stress level and decreases with lowering the temperature.

During creep under deviatoric stress in frozen soil the creep rate first decreases up to a certain point, and then increases again until finally failure occurs. In the literature there are usually three creep stages defined, i.e. primary creep with decreasing creep rate, secondary creep with constant creep rate and tertiary creep with increasing creep rate. Theoretical investigation as well as high precision tests however indicate that a secondary creep phase with constant creep rate does not exist. Furthermore, even under very low deviatoric stress, the stage with increasing creep rate is eventually obtained. So, frozen soil is rather a fluid of high viscosity.

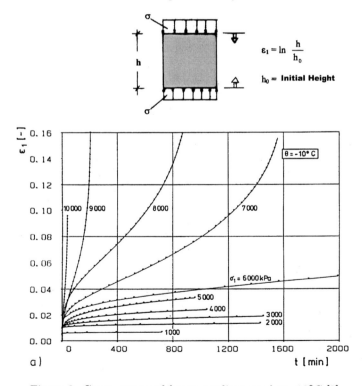

Figure 3: Creep curves of frozen medium sand at -10°C [3]

Tensile strength of frozen soil is approximately $1/5 - 1/4$ of its compression strength. Triaxial stress states can increase the strength by activation of friction, on the other hand mainly in highly saturated fine grained soils it can reduce their strength by melting of the ice matrix.

Frost heave occurs by the expansion of water during freezing, if the soil is fine grained and the water cannot escape. Further, mainly in fine grained soils and a slow frost propagation, water is attracted to the freezing front and locally expells the soil so that ice lenses are created. The amount and velocity of ice lense creation is highly dependent on the frost propagation rate, the permeability of soil and the overburden pressure.

In frost heave susceptible soil, settlements occur during thawing, that can be larger than the original frost heave.

Chemical composition of the water, mainly salt, can significantly affect the mechanical behaviour of frozen soil.

4   Examples of ground freezing in tunnelling

4.1   Longitudinal freezing pipes

In the city of Fürth a 90 m long section of a subway tunnel, starting out of a shaft, was driven. Longitudinal freezing pipes were drilled at lengths up to 60 m to freeze a shell above the tunnel. Normally, horizontal drilling is limited to a length of approximately 25-30 m because otherwise the deviation from the planned position is too high. Therefore a special flush drilling system was installed, that allowed checking, steering and assessing the drilling accuracy.

The upper part of the cross section of the tunnel was situated in a quarternary sand layer, the bottom of the cross section was in a nearly watertight sand-stone. The top of the sandstone ascended within approximately 60 m to the tunnel top, so the frozen shell could end there, see Fig. 4.

Special regard was necessary to the waterflow from one side, that delayed the frost propagation. Calculations carried out in advance indicated, that closing the frost wall with only one row of freezing pipes would not be possible. As there where three rows of freezing pipes around the two tunnel tubes, cooling of groundwater at the first and the second row was taken in account. The calculation showed that it would be possible to close the freeze wall at the third row with this pre-cooled groundwater. The measurement of frost propagation during freezing indicated indeed that the frozen wall was closed

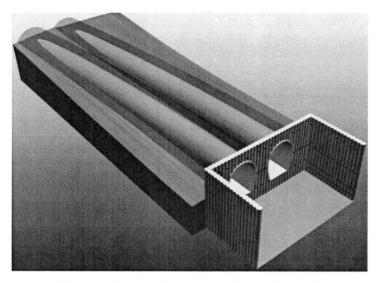

Figure 4: Frozen soil shell above the sandstone [1]

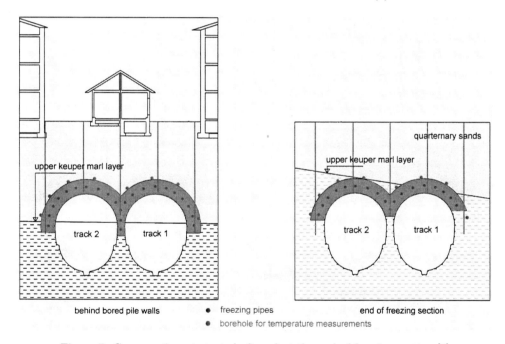

Figure 5: Cross section at start shaft and at the end of freezing section [1]

at the third row first and a short time later the frozen body at the second and the first row of freezing pipes were closed.

The freezing pipes at the full length of the tunnel allowed a fast driving without interrupt. However it has to be taken in account that the freezing energy is constant over the whole length of the tunnel, so freezing increasingly crosses the working face. Therefore, in the beginning the working face was nearly unfrozen but in the later stages the frost had penetrated the tunnel more and more, and this fact delayed the excavation of the soil.

4.2 Staggered freezing pipes

As a part of the highway system of the city of Zürich / Switzerland the Milchbuck Tunnel was constructed by use of ground freezing in the southern moraine section, se Fig. 6. The tunnel had a comparatively large cross section of 145 m^2 and the distance to the buildings above is partly only 5.5 m. The 350 m long moraine section was subdivided in freezing sections of 30-40 m. This length resulted from the requirement of a thin frozen ground arch and the accuracy of the horizontal drill holes.

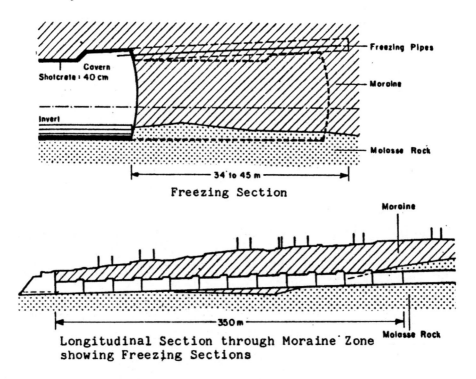

Figure 6: Longitudinal section Milchbuck tunnel [6]

The normal cross section was enlarged at the end of each freezing section to 195 m^2 over 8 m length. From these drilling chambers the freezing pipes have been drilled approximately 4-6 m ahead of the end of each section. The pipes had an exterior diameter of 140 mm, the casing of the boring was used as freezing pipe and was left in the ground. The spacing between the freezing pipes was 1.0-1.25 m.

The brine temperature was first -20°C, for the third and the following sections it was changed to -35°C. As the freezing pipes are closer at the beginning of each section, this zone froze at an earlier time, and this permitted to start the excavation already 5 to 7 days after starting the freezing operation. The growth of the frozen body and consequently also the frost heave was limited by intermittent freezing. Together with a shorter duration of the freezing in the later sections this let to a reduction of the maximum frost heave of 105 mm in the first freezing section to a few millimeters later.

The use of staggered freezing pipes in tunnelling reduces the excavation rate because of the interruptions for drilling the next freezing pipe section. On the other hand, the growth of the frozen body can be controlled very well to prevent frost heave and freezing of the working face.

4.3 Freezing from a pilot adit

In the city of Munich directly below the townhall the existing subway station was to be enlarged to accept the increasing number of passengers. The freezing method was adopted. The freezing pipes were drilled from a pilot adit above the tunnel profile. The originally planed solution was a NATM tunnel drive at each side of the subway station and a row of horizontally grouting pipes above the tunnel to compensate settlements of the townhall.

The soil existed of frost susceptible silt and clay directly above the tunnels and fine sand layers. There where two groundwater levels, which where originally separated, but likely with conjunctions because of well drilling at former construction activities. The position as well as the filling of these wells where no more known.

According to the contractors (Bögl company) proposal, two pilot adits were driven from two starting shafts under compressed air at a length of approximately 100 m, see Fig. 7. They ended in the diaphragm wall of the existing subway station Marienplatz. From each of the two pilot adits approximately 2000 m freezing pipes were drilled into the soil. This had to be done against groundwater pressure. With these freezing pipes the frozen body was created.

The frost heave behaviour was investigated before in tests at the Universit'e Laval at Quebec/Canada. Based on these tests, the maximum frost heave was predicted to approximately 15 mm at the footings of the townhall. As there were a lot of at settlement measurements taken out by the Munich Technical University, the behaviour of the frozen body could be observed very precisely.

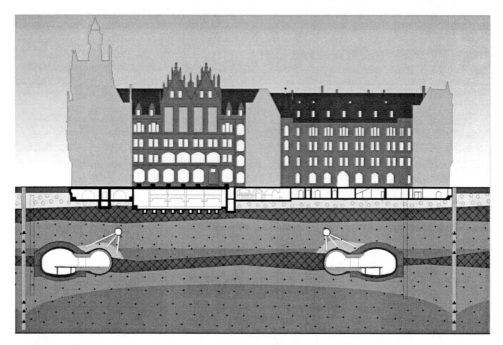

Figure 7: Cross section of the subway station Marienplatz in Munich with pilot adits and freezing bodies [[2]

The maximum measured frost heave was 12 mm and thus was within the limited range. The maximum thaw settlement was approximately 17 mm, what resulted in net settlements of approximately 5 mm.

The excavation of the tunnel was done without major problems, at two positions small water penetration were encountered because of unknown materials brought into the ground during the construction of the existing tunnel. These spots could be closed by additional freezing, so they did not significantly delay the driving of the tunnel.

4.4 Soil freezing for jacking a tunnel frame under a railway embankment

This construction method is not a classical tunnelling method, nevertheless it leads to a traffic passage without cutting the ground surface. The purpose is the construction of a new road underpassing a railway line. As this track is passed by 330 trains per day and a diversion does not exist, it had to be kept in service. The allowed maximum interruption time was 8 hours at weekend and 20 or 30 minutes at day and night, respectively, in working days.

As the road level needed to be as high as possible because of ground water, a 2600 t heavy concrete frame was constructed beside the embankment and then jacked through it by a hydraulic system. The tracks were kept up by steel girders which were supported dependent on their position on the concrete frame and on the other side on the earth embankment. As the railway embankment consisted of 3 m loose filled soil overlying sandy gravel with some silt below the working face, it was not stable under the load of the passing trains. To improve its stability, it was planned to drive vertical piles of wood or steel from the embankment surface to carry the train loads to deeper layers. This was a very time-consuming method in view of the short working periods between train passages. Further, if any difficulties were to arise during driving a pile it would be necessary to interupt it when a train approaches and restart with a new pile after the train has passed.

The main disadvantage of the piling system however was that the slope had to be cut vertically to remove the piles. This step was very critical and experience showed that neighbour piles could drop out immediately after removing the first pile of a row. It then could become necessary to stop all trains until the concrete frame was pressed close to the new embankment crown supporting the steel girder again.

To avoid these disadvantages, a horizontal frozen slab was installed directly below the tracks respectively the steel girder. This slab transferred the track load away from the slope edge to the backwards area and distributed the load in the track direction increasing thus the slope stability. Furthermore, the frozen slab damped the dynamic load from the trains.

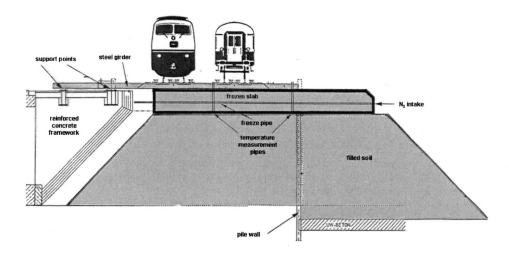

Figure 8: Cross section with frozen slab

During jacking of the concrete framework, the frozen slab had to be cut off. This was done by blasting with holes drilled from the surface. As liquid nitrogen was fed from one side and the gas outlet was on the other side at the working face, the comparatively soft copper pipes were cut together with the soil. The freezing pipes had a spacing of 0.9 m and were approximately 24 m long in the beginning.

To control liquid nitrogen intake, a thermo-couple was placed in every freeze pipe near the outlet. These thermo-couples were torn back with cables to the intake side before cutting the next section of the pipe. To control the frozen body's thickness, and temperature additionally six vertical measurement pipes had been driven from the embankment surface into the later frozen area with three thermo-couples in each of them.

The frost heave reached a maximum of 2.2 cm within the lifetime of one week and was distributed nearly equally over the frozen area. This was more convenient than the peaks and valleys that usually are created by the support points above piles. At both sides the tracks were ramped by wedges so train passengers could not feel any shocks when the trains passed the building site.

References

[1] Bayer, F.: Baugrundvereisung beim Bau der U-Bahn Fürth / (*Subsoil Freezing during the building of the Fürth underground*), Tunnel 7/2002, pp. 20-28

[2] Müller, B., Selmer, K.: Baugrundvereisung am U-Bahnhof Marienplatz in München / (*Subsurface freezing at the Marienplatz underground station in Munich*), Tunnel 1/2005, pp. 30-36

[3] Orth, W.: Gefrorener Sand als Werkstoff – Elementversuche und Materialmodell / (*Frozen soil as material – element tests and constitutive model*), Publications of the Insitute of Soil and Rock Mechanics, University of Karlsruhe, Vol. 100, 1986

[4] Orth, W.: Frostkörperstatik und thermische Berechnungen, U-Bahn Fürth, Strecke Klinikum bis Bhf Stadthalle / (*Statics of frozen bodies and thermal calculations, Fürth subway line, section Klinikum - Stadthalle*), 2000, (unpuplished)

[5] Rögener, B., Orth, W., Steinhagen, P.: Durchpressung einer Eisenbahnüberführung mit Vereisung im Zuge der Ausbau- und Neubaustrecke Karlsruhe-Basel, Bauingenieur 68 (1993), pp. 451-460, Springer-Verlag, Berlin

[6] Schmid, L.: Milchbuck tunnel – Application of the freezing method to drive a three-lane highway tunnel close to the surface. RETC Proceedings Vol. 1, Society for Mining Metallurgy, pp. 427-445, 1981

The start of hydroshields - risk assessment under high water pressure

Jürgen Schwarz

Dywidag Bau GmbH, Aschheim bei München, Germany

Abstract: The subject of this article is tunnelling under a high groundwater level in loose soil, especially in gravel and sand. Shortly I will describe suitable tunnelling machines and the technique to build the shafts for start and end of the tunnelling process.

The focus of the article is the intersection of start shaft and tunnelling process, i.e. the moment when the tunnelling machine leaves the start shaft.

1 Tunnelling machine

During excavation process in sub soil under ground water the machine has to support the ground, it has to ensure equilibrium with water pressure and earth pressure.

The most critical point is the face of the excavation. The cutterhead excavates the ground, the support of the face can be achieved by the following two methods:

- earth-pressure-balance (EPB)

- slurry

The range of application for these techniques is shown in Fig. 1 and 2. [1]

The EPB-machines are used in fine soils as clay and silt, the slurry machines in sandy and gravelly soils.

1.1 EPB-machines

The EPB-principle is: the soil loosened by the cutting wheel serves to support the working face. A screw conveyer transports the soil out of the chamber on a conveyer band (Fig. 3).

The suitable soil for this system has

- low consistency

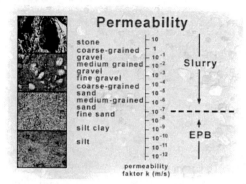

Figure 1: Range of application of slurry and EPB technique in dependence of permeability

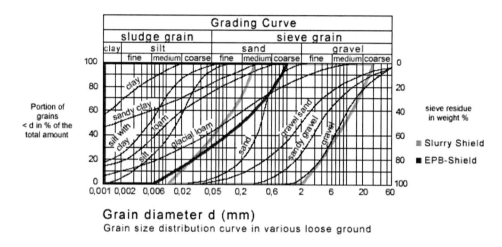

Figure 2: Range of application of slurry and EPB techniques in dependence of grain size

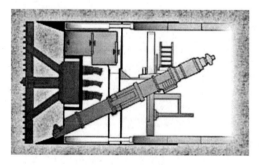

Figure 3: EPB shield

- high plasticity

- low friction

- low permeability.

It is normally a clay or silt, but using polyurethane foams also gravelly soils can be conditioned for an EPB-shield.

1.2 Slurry-Shield

The slurry-support works on the basis of a slurry-mix of bentonite and water which is pumped into the excavation chamber. A slurry reservoir and a cushion of compressed air behind a submerged wall guarantee the sufficient pressure at the face (Fig. 4).

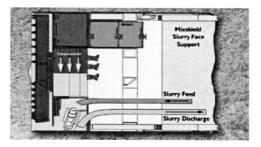

Figure 4: Slurry shield (mix shield)

The soil-slurry-water-mix is pumped via the discharge pipe to the separator outside the tunnel. Fine soils must be separated before bentonite can be recycled.

The separation costs grow exponentially when fine soils are excavated. Therefore the use of slurry shields is normally limited on sandy/gravelly soils.

2 Shafts

A shaft in subsoil under high water level consists of

- watertight walls

- watertight bottom

The usually applied techniques for the construction of watertight walls are

- diaphragm walls (Fig. 5)

- bored pile walls (Fig. 6)

- sheet pile walls (Fig. 7)

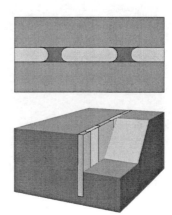

Figure 5: Diaphragm wall

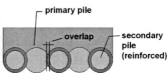

Figure 6: Bored pile wall

Figure 7: Sheet pile

In addition, there are further possibilities using injection technologies.

For the watertight bottom can be used a watertight soil layer, if it exists in a sufficient depth (for uplift safety) but not too deep (for economic reasons).

Artificial bottom solutions are (Fig. 8)

- underwater concrete (tied with vertical anchors acting against uplift)

- injection bottoms

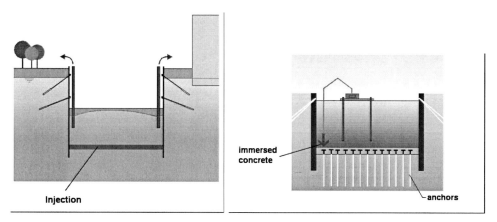

Figure 8: Watertight bottoms of shafts

3 Machine start

The start of a shield machine is an evident example for the principle of tunnelling under high water level. On one side — inside the starting shaft — prevail atmospheric conditions. On the other side — in the ground — acts hydrostatic pore water pressure groundwater level. At every time, at every step of the procedure at least two independent sealing bodies must be guaranteed.

Sealing blocks behind the lining of the shaft improve the intersection.

3.1 Requirements on the sealing at the start shaft

The requirements on sealing depend on the soil and the groundwater conditions. In Fig. 9 is shown a situation which is typical in Munich:

- well graded gravel, homogeneous soil

- moderate water pressure

In this case an improvement of the soil adjacent to the start shaft would be sufficient. Any additional sealing measure is not necessary, as any incoming waters can easily pumped off the start shaft. Erosion and related progressively increasing water discharges will not occur.

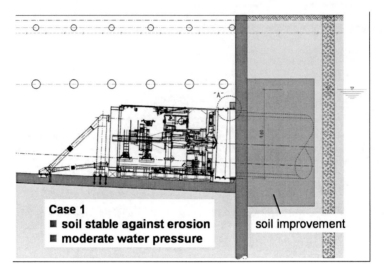

Figure 9: Drive-in operation: requirement of a soil improvement

In the second case (Fig. 10), a veritable sealing body is necessary. This example stems from the construction site metro Rotterdam/Statenweg (joint venture Walter Bau, Dywidag, Züblin, Duravermeer). The design was by a consortium of Walter Dywidag Engineering, Züblin TBT and WD Engineering. The geological situation is characterized by

- inhomogeneous and fine soil, susceptible to erosion

- high water pressure.

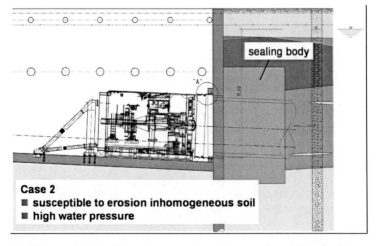

Figure 10: Drive-in operation: requirement of a sealing body

The requirements on the sealing body are considerably higher than in case 1. Here was needed not only a soil improvement but also a veritable sealing body. How hazardous such a situation can be is shown in Fig. 11. Minute percolations starting with ingressing droplets can very soon leed to a veritable catastrophy [2].

Figure 11: Rescue of a tunnel excavator after a catastrophy; Leidingen tunnel, Rotterdam (1998)

The state of the art is to permanently hold two independent redundant sealing measures. The second sealing measure is a leak-proof frame adjacent to the improved soil. This frame is twofold composed from a lip seal and an emergency seal. There are further developments of the leak-proof frame with respect to their sealing profiles and extensions up to a 'sealing pot' [3] or a starting chamber, but they are not included in this paper.

3.2 Phase 1: Break-open the face wall

In the first phase of the starting operation, the machine must break-open the face wall. Usually the face wall (e.g. a diaphragm wall) is reinforced, and the machine cannot cut the reinforcement. Thus, the reinforcement has to by previously cut by hand. During this stage the only sealing body is the improved soil behind the wall.

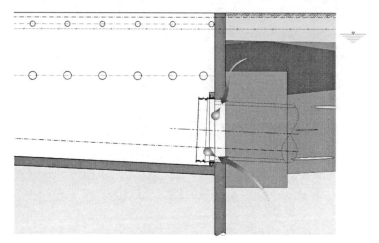

Figure 12: Break-open the face wall

The risks are:

1. The connexion of the sealing body with the wall

   Joints are always critical! The connexion between sealing body (improved soil) and face wall is even more critical, if the excavation of the shaft occurs after the soil improvement, because then deformations of the wall will detach it from the adjacent improved soil.

2. The sealing body itself

   The usual technique to improve the soil is jet grouting. Despite all technological advances and controls, the resulting material is not sufficiently homogeneous and resistent against erosion. Improved jet grouting techniques cannot help in inhomogeneous soils, as e.g. the one in Rotterdam. The ranges of coverage and the strengths of the obtained soil are to such an extend variable (in sand and gravel different than in silt or clay) that the 'improved' soil is not homogeneous and has a layered struture (like filo pastry). A spectacular failure of such a sealing body occured at the Fernbahn-tunnel in Berlin.

   The second sealing measure must be available from the very beginning of the starting operation. A possibility is to break-open the face wall under the protection of pressurized air, which prevents inrush of water. It is convenient to use the shield as pressurized chamber. The shield itself is equipped with an air-lock. With appropriate design of the pressurized chamber, there will be sufficient work space in front of the cutting wheel. After breaking the wall, the starting operation can begin. The leak-proof frame provides a sufficient sealing owing to the waterproofing performance of the seal-lips at pressurized air. The shield is placed in such a way that it fits to the leak-proof frame and, during this phase, is not being moved.

A new variant is to use glass fibre reinforced plastic bars instead of steel rebars to reinforce the face wall. In this case the cutterhead is equipped with disc cutters to penetrate the face wall.

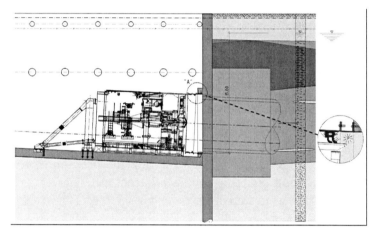

Figure 13: Break-open the face wall under the protection of pressurized air

3.3 Phase 2 - Penetration into the sealing body

In this phase the leak-proof frame must stop any water ingress due to the aforementioned voids in the sealing body. The conventional lip seal is shown in Fig. 14. It can tolerate deviations up to 50 mm. Tolerance and conicity of the shield take 20 mm out of these 50 mm. The remaining 30 mm are consumed by an angular deviation of 4.5° in the case of a shield ∅ 9 m. During the passage of a, say, 10 m long shield, this angular tolerance must not be exceeded.

In case of high water pressure and/or erodible soil an improved sealing is needed. This can be achieved by building a coffer adjacent to the face wall and lowering the watertable within this coffer (Fig. 15).

An additional increase of safety can be achieved by filling this caisson with immersed concrete, which is considerably more homogeneous, dense and resistent to erosion than jet-grouted soil. During/after passage of the shield the

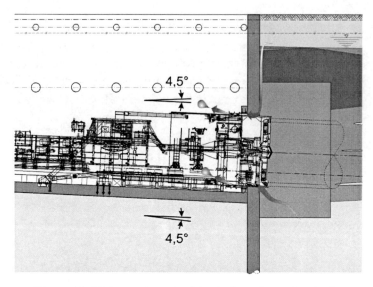

Figure 14: Risks at phase 2: drive-in into the sealing body

sealing is achieved by the tail-gap grouting. This is the safest but also the most expensive way of sealing. The coffer can also be built with a sheet pile wall, but the sheet piles must be removed for the drive-in operation. It is difficult to remove (pull out) sheet piles adjacent to concrete or jet-grouted soil. In addition, this would create a deficient sealing. Blow-outs could then occur at this weak point.

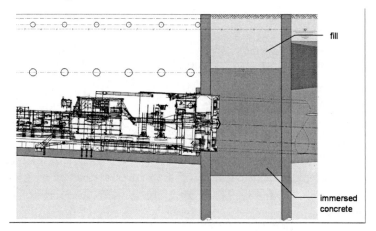

Figure 15: Immersed concrete sealing body

3.4 Phase 3 - Drive out from the sealing body

With whatsoever construction of the coffer, the increase of safety does only pay, if this can be achieved for a sufficiently long section. This means that the lining segments must have reached the leak-proof frame when the shield leaves the coffer. Only then the limited tolerance of the leak-proof frame is no more a problem. The grouted tail gap creates then a permanent sealing (Fig. 16).

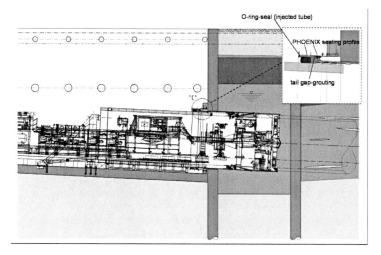

Figure 16: Long coffer

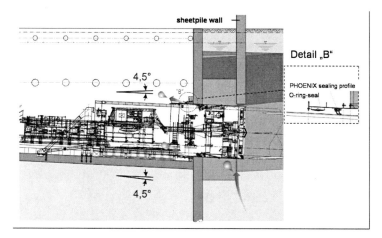

Figure 17: Short coffer with sheet pile wall

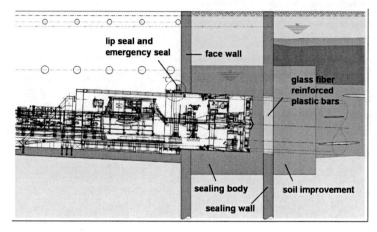

Figure 18: Sealing measures for leaving the start shaft

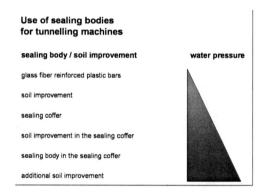

Figure 19: Methods of soil improvement as function of water pressure

4   Conclusion

Tunnelling under high water pressure is a problem of intersection

- between face and soil
- between start shaft (wall and bottom) and soil
- between start shaft and tunnel.

The main principle is: we need at least two independent sealing systems at any place and at any time. This principle was described in detail for the starting process. It is the duty of the engineer to guarantee the two independent safety barriers for the entire tunnelling process.

References

[1] Publications of Herrenknecht AG, Germany

[2] Schwarz, J. *et al.*: Lösungen für dem Anfahrvorgang von Hydroschilden unter hohem Wasserdruck (*Solutions for the drive-in operation of hydro shields under high water pressures*, in German, Unterirdisches Bauen 2000 – Herausforderungen und Entwicklungspotentiale, Forschung und Praxis 38, STUVA-Tagung 1999)

[3] Schwarz, J. *et al.*: Realization of new tunnelling methods in Metro Amsterdam Nord/Zuidlijn – The contractors point of view, ITA World Tunnelling Congress, Amsterdam 2003

Geotechnical risks for tunnel drives with shield machines

Markus Thewes

STUVA – Studiengesellschaft für unterirdische Verkehrsanlagen e.V., Mathias-Brüggen-Str. 41, Cologne, Germany

Abstract: Focussing on tunnel drives with shield machines like EPB- or Slurry-Shields in soft ground, critical geotechnical risks with essential importance to the success of a tunnel project are identified. Reflecting on the continuing efforts to build larger, longer and deeper tunnels in increasingly difficult ground conditions, the available methods for risk mitigation are discussed regarding their advantages and the limits of their application.

1   Introduction

During tunnel drives with shield machines the expert knowledge and the experience of geotechnical engineers in many ways is of essential importance for the success of a tunnel project. This does not only apply to questions dealing with the influence of the ground on the structural stability of the shield machine and the structural stability of the tunnel lining. It also applies to geotechnical analyses and evaluations which are important for the operation of a shield machine.

In this paper some of the connections of geotechnical risks and practical solutions are presented, using the examples of several recent tunnel drives.

2   Geotechnical risks for tunnel drives with shield machines

Recent experience shows that detailed and thorough evaluations of all risks related to tunnel works are of essential importance to the technical and economical success of a tunnel project. Accordingly, clients and insurers have started to demand risk evaluations in the early phase of tunnel projects [1].

The geotechnical risk issues which have to be discussed for a tunnel drive with a shield machine in detail are related to the following fundamental areas:

- Structural design of the shield machine

- Structural design of the tunnel lining

- Stability of the tunnel face and safety of workers at the tunnel face

- Minimizing the influence of settlements on existing structures

- Control of the ground water during construction and operation

- Protection of workers and environment in handling of ground water and soil, esp. when contaminated

- Ensuring a sustainable and durable design of the tunnel lining including its seals for a time span of operation, which usually exceeds 100 years.

Also, geotechnical engineers supply an important basis of decision-making for the choice of tunnelling technology to be applied:

- Ensuring high daily advancement rates by making the right choice for the type of shield machine

- Ensuring a controlled support of the tunnel face

- Ensuring a reliable cutting process of the ground with minimized wear of the cutting tools and the cutter head

- Ensuring a successful mucking process from the excavation chamber of a shield, to the separation and handling of muck on the jobsite, to the depositing or the re-use of the soil in other applications.

The particular geotechnical risks of each tunnel drive have a strong influence on the equipment and the cost of the shield machine to be used. Some of these risks can only efficiently be mastered by applying recent developments in shield technology. Some of the practical developments achieved are partly quite unusual and will be described to more detail in this paper:

- Adhesion of clay leading to clogging of the excavation chamber

- Accessing the excavation chamber under very high support pressures which exceed the suitability of compressed air works

- Transition zones with mixed face conditions of soil and hard rock leading to extreme demands on the technology for cutting the ground and supporting the tunnel face.

For the sake of completeness, other geotechnical risks should be mentioned in this context, which will not be discussed in greater detail within this paper. They, however, need to be identified and discussed during geotechnical analyses for shield drives, so that they can be included in the concepts for the installed technology and for the operation of a shield machine:

- Low overburden: Increased risk of settlements, limited access to the excavation chamber in ground conditions with ground water

- Particularities of the grain structure of sand and gravel: Strong dilatancy leads to underpressures in the pore water during the cutting process, increasing the effective stresses and thereby increasing tool wear

- Very coarse grained, highly permeable soil with ground water: Face support becomes very difficult

- High content of fines in silty and clayey soils: Increased effort for separation of the bentonite suspension when using a slurry shield

- Swelling soil or rock: Increased ground load on the shield machine and on the lining becomes possible; longitudinal water drainage in the grouted annular gap should be avoided

- Contaminated soil: Muck and possibly also the supporting fluid of a slurry shield may be contaminated; contamination may be spread further by bentonite slurry

- Man-made objects in the ground: Mostly in urban areas anchors, sheet pile walls, and piles can lead to problems

- Close proximity to existing structures in the ground: Other tunnels, foundations, and sewer lines may be endangered by stress redistribution and by settlements induced by a tunnel drive.

3 Clogging in clay formations

During tunnel drives in clay formations with shield machines with a fluid-supported tunnel face extensive clogging may occur (figure 1) leading to difficulties for the excavation process. At first, the clogging hinders the transport of the spoil in the cutting wheel, excavation and suction inlet area and then leading to further obstructions which can result in a complete interruption of the material flow. If clogging cannot successfully be mitigated, it can lead to a drastic performance reduction due to a reduced advance rate and to the time required for additional cleaning efforts.

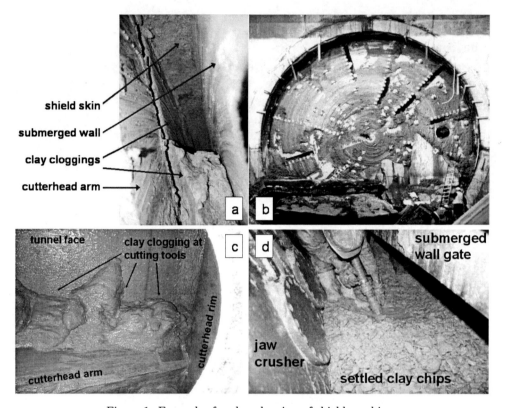

Figure 1: Examples for clay clogging of shield machines

3.1 Development from first clogging to standstill

In practice, two developments can be observed from the first appearance of clogging to a complete blockage, stemming from the two areas mainly responsible for clogging - the cutting wheel and the suction inlet area. These mechanisms can also be overlapping.

Often the problem first occurs in the suction inlet area due to a build-up of material in front of the inlet grill. An accumulation of excavated clay fills up the bottom area (figure 1) and is then compressed by further spoil. In a short time, the suction inlet area and the excavation chamber can be filled with clay. The compaction of the material in the rear working chamber can even be increased by the cyclic movements of a jaw crusher, pushing the spoil which has settled in the bottom area up towards the side.

In the second case, clogging occurs in the cutting wheel area (figure 1 a, b) there is first a build-up on the cutting tools, particularly in the cutting wheel centre. As a consequence, clay discs typically occur in front of the cutting wheel. This applies, in particular, to the centre areas of cutting wheels not provided with an active centre cutter, as the excavated material cannot directly flow from the cutting tool towards the excavation chamber but has to flow parallel to the tunnel face over a certain distance. The effect of the clogged centre area of spoke-type cutting wheels (and also for the entire surface area of closed-type cutting wheels) is that the soil is displaced instead of being cut. This can lead to high contact pressures of the cutting wheel resulting in an increase of the torque and stalling in the worst case.

A mixed tunnel face with clay and gravel or clay and sand can lead to additional stability problems of the tunnel face, as the required fluid support is no longer provided when the excavation chamber is clogged. Furthermore, the abrasive paste may lead to a high wear, in particular in the area of the gauge cutters around the periphery due to the great speed differential between the rotating cutting wheel and the stationary cutting edge of the shield.

3.2 Clogging potential as a sum of four single effect mechanisms

In order to classify soils clay formations regarding their clogging potential, a research program on the adhesion of clay on steel has been carried out [2]. The program was based on practical research regarding the clogging potential as well as laboratory tests with clay samples. The clogging potential of a clay formation was defined as the interaction of four single effect mechanisms (figure 2):

- the adhesion of clay particles on a component surface,
- the bridging of a clay particles over openings in the path of the spoil transport
- the cohesion of clay particles, sticking to each other,
- the low tendency of a clay towards dissolving in a water.

From these four mechanisms, adhesion is the most important mechanism and characterizes best the sticky character of clay. Depending on the relative movement between soil and steel it can be subdivided into normal and tangential adhesion.

3.3 Laboratory test to define clay adhesion

A laboratory test has been developed in order to evaluate the adhesion between the soil sample and a steel plate when it is pulled vertically from the

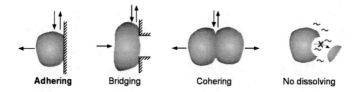

Figure 2: Clogging potential – interaction of single effect mechanisms

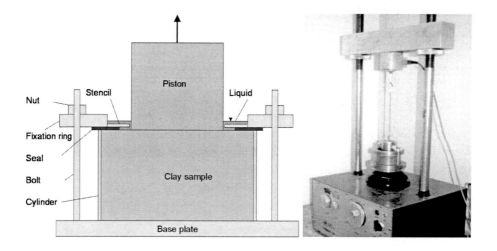

Figure 3: Schematic diagram of separation test for clay adhesion, photo of test setup

sample (figure 3). Wetting (with a fluid similar to the support fluid of a fluid-supported shield) is vital prior to soil and steel contact.

The adhesion tests have been carried out with samples from six clay formations varying in their mineralogical composition. The normal adhesion of soil and steel proved to depend strongly on the content of swelling clay minerals (illite, montmorillonite) and the consistency of the soil.

The test results allowed the development of a model for the principle mechanism during the adhesion of swelling clay as shown in schematic sketches in figure 4. On the left side, the sketch shows the wetting of a soil sample with water under atmospheric pressure. The pore pressure in the swelling clay is much lower than in the water. The resulting pressure difference causes the clay to absorb the water and the clay swells.

On the right side of figure 4, a steel plate is shown which has been wetted

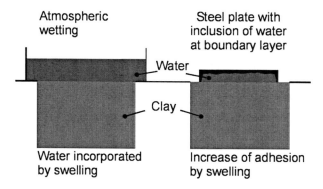

Figure 4: Schematic diagram on the model for adhesion of a swelling clay formation

prior to soil contact. As even a polished steel plate is rough in microscopic terms, there is a small amount of water enclosed between the soil and the steel surface. The clay, like in the first example, tries to absorb the enclosed water. After a certain period of time, as no water can flow to be absorbed by the clay, a subpressure with a value similar to the pore water pressure of the clay develops in the water, which is encapsulated at the clay-steel interface. The result is a tensile stress at the interface (adhesion).

During the tests, adhesion values of up to 208 kN/m^2 have been measured. This compares a column of clay of approximately 11 m length, which is "suspended" to a steel interface by adhesion!

3.4   Identifying and classifying soils with clogging potential

During the research program a broad practical study was carried out regarding soils with clogging potential. The geological information from many tunnel drives, where problems of different scales occurred due to clogging, was analyzed. During these projects conventional fluid-supported shields were used which originally were not designed for cohesive clay soils. The evaluation of this data leads to a classification of the soils into three categories:

- Soils with high clogging potential lead to substantial problems during excavation and required daily cleaning works. The modifications, which were subsequently installed underground, only led to a reduction but not a solution of the problem.

- Soils with medium clogging potential could be mastered after a number of mechanical modifications of the shield machine and its slurry circuit along with changes in the operation of the machine (reduced advance rate, longer flushing periods).

- Soils with low clogging potential required a reduction in the advance rate and additional flushing periods, but could be overcome only by reducing the excavation speed of the TBM without making major alterations of the machinery.

In all three categories information on grain size distribution, consistency limits and the natural moisture contents of the excavated clay was available. Figure 5 shows the relationship between the consistency and plasticity indices and the categories for the clogging potential. This diagram provides an indication of clogging risks for fluid-supported shield drives.

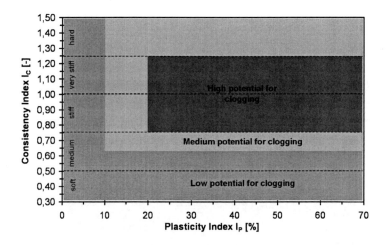

Figure 5: Clogging potential of clay formations regarding plasticity and consistency

3.5   Preventive measures against clogging

For all fluid-supported shield tunnel projects where a significant amount of clay has to be excavated, it is recommended to consider preventive measures in terms of the design of the tunnelling system and the separation plant, even if the clay only shows a low potential for clogging. The principle measures in the design of the shield machine to prevent clogging generally are as follows:

- Generate large soil chips to reduce the adhesion-prone surface of the excavated clay in relation to its volume.

- Avoid narrow passages and other obstructions for the transport of a clay chip from the tunnel face to the slurry line

- Avoid clay accumulation and minimize the time spent by the clay chips in the chamber by increasing the ratio of the suspension flow rate to the volume of excavated soil.

- Avoid clay agglomerations through increased agitation in areas which are prone to material settlement.

For clays with low to medium clogging potential the risk of adhesions generally is low, whereas a risk of clogged openings still remains during spoil transport, depending on the consistency of the material (bridging). In this regard, narrow passages should be avoided. Unfavorable elements in terms of design are the closed-type cutting wheels, the centre area of spoke-type cutting wheels and grills, used in front of the suction inlet.

A clay with medium to high clogging potential can adhere to all parts of the machine, to which it comes into contact with over a sufficient length of time. In order to use a fluid-supported shield in such a soil, a consequent and costly optimization is required regarding the entire excavation and transport process:

- Narrow passages (on the cutting wheel) must be avoided,

- the soil flow should not lead to direct impacts on components (cutting tools, sieve, stone crusher),

- 90° angles in the corners of the excavation chamber must be broken or rounded,

- in areas where material may settle turbulence must be created (flushing nozzles, agitators) and

- the flow rate of the suspension in the excavation chamber must be maximized (circuit and flushing concept).

3.6 Operational measures against clogging

Once clogging occurs in fluid-supported shield drives, there remain but few operational measures in order to restrict the resulting problems:

- A drastic reduction of the advance rate

- An optimization of the cutting tool penetration in order to get a favorable size of the clay chips

- A maximum suspension flow rate with a feeding system directly at the cutting wheel or in the excavation chamber

- Flushing pauses or flushing of the excavation chamber during ring-building

- Operation with a partial air support in the excavation chamber (only possible with a stable tunnel face)

- The use of chemicals as additives.

3.7   Experience from a successful tunnel drive in very adhesive clay

The Westerschelde tunnel in the Netherlands was a pioneering project in terms of the geotechnical evaluation of the clogging potential of the soils to be encountered and the design of the shield machine for the prevention of clogging. The Boom Clay a tertiary, overconsolidated, highly plastic, illitic-montmorillonitic and thus strongly swelling clay was encountered over two thirds of the total length of the tunnel ($2 \times 6,600$ m, figure 6). The success of the project was closely linked to the possibility of a trouble-free advance through the Boom Clay.

Identifying the high clogging potential of the Boom Clay led to a consequent optimization of the cutting wheel, the excavation chamber, the suction inlet areas of both mixshields, manufactured by Herrenknecht AG [3]. Furthermore, the clogging potential has been taken into account during implementation of a sophisticated flushing system in the excavation chamber and design of the data acquisition system [4].

The cutting wheel (figure 7) was designed as open-spoke cutting wheel with rim and the form of the cutting arms was streamlined. Soft-ground cutting tools were used, allowing for the excavation of isolated clay chips. A centre cutter with a diameter of 2.5 m was installed on the axis of the machine, equipped with individual feed and slurry lines. The excavation chamber was lined with steel sheets to avoid clogging-prone angles.

In the crusher chamber rotary crushers with two agitators were installed (figures 7 and 8). These crushers were able to break larger lumps of the hard clay to an adequate size for pumping. The chosen combination of rotary crushers

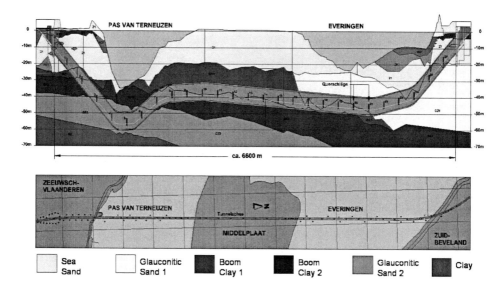

Figure 6: Longitudinal section and plane view of the Westerscheldetunnel

Figure 7: Herrenknecht shield machine with detail of the suction inlet area

and agitators did not restrict the slurry flow in the same way as a combination of jaw crusher and sieve would have.

In the excavation chamber and the suction inlet area an innovative flushing system was implemented (figure 8). The feed flow of max 2,200 m^3/h could be distributed individually to various parts of the excavation chamber and the suction inlet area.

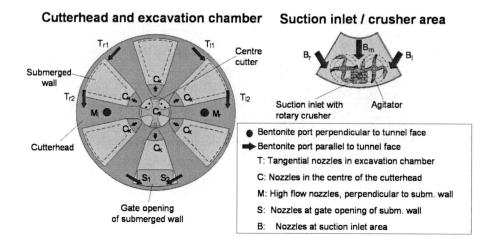

Figure 8: Scheme of the flushing circuit of the Herrenknecht TBM

Both tunnel boring machines have successfully excavated the Westerschelde-tunnel including the sections in the Boom Clay. Despite the very high clogging potential of the Boom Clay, the tunnelling works did not come to a stop due to clogging in any part of the tunnel drives. Using the flushing system an excavation chamber, which was clogged by about 70% could still be cleaned by the machine itself.

3.8   Geotechnical studies for future shield drives

Problems in shield driving caused by clogging can be avoided most efficiently when an existing clogging potential is detected beforehand and the tunnel boring equipment as well as the estimated performance rate are adapted correspondingly. The geological expertises for future construction works with fluid-supported shields should take this aspect into account. The geotechnical evaluation of cohesive soils thus should consider the following requirements:

- Evaluation of the plasticity number and the consistency number of all relevant clays to indicate the clogging potential according to figure 5

- Clay mineral analyses to define the mass fraction of the most important minerals (caolin, illite, montmorillonite, smectite, quartz etc). Starting from a content of swelling clay minerals (illite, montmorillonite) above 10%, adhesion must be taken into account.

- More detailed information regarding clay formations to evaluate more accurately the length of the tunnel and amount the tunnel face affected.

3.9 Adhesion tests in practice

Unfortunately, little practical information on the relation between adhesion tests and practical tunnelling experience has been published at present. The results of various adhesion tests developed by different geotechnical laboratories cannot be compared to each other because of differences in the testing procedures. Therefore, bidding contractors cannot evaluate the results of adhesion tests in construction practice, yet. This will be the case as long as there will be no unified geotechnical test procedure allowing extensive testing of characteristic soils, which are known for their clogging potential in tunnelling.

Until then, clients, engineers and contractors will at least be able to work with the qualitative indication of the clogging potential as described in this paper.

3.10 Recommendation for risk analyses

The economic success of a tunnelling project can depend very much on the clogging potential of the encountered clay formations. The production progress in clay can be 2/3 (see above) or possibly as low as 1/10 of the regular progress in granular soil. The clogging potential thus represents a decisive characteristic with high influence on the performance of tunnelling works with fluid-supported shield machines. A careful evaluation of the clogging potential and the adhesion of clays to be encountered as well as a good quantification of the affected length of a tunnel can be helpful in establishing more contractual transparency and in avoiding disputes in the construction stage.

4 Maintenance work in the excavation chamber under high pressure – "Diving in a shield machine"

In situations with unstable ground, if maintenance is to take place in the excavation chamber of a shield machine with active face support, the tunnel face has to be supported with compressed air. The slurry shield technology is especially well suited for this because the bentonite suspension used for face support and for the transport of muck from the excavation chamber leaves a membrane-like bentonite layer on the surface of permeable soils, which reduces the loss of compressed air into the ground.

At the Westerscheldetunnel (see fig. 6), the deepest section of the tunnel
occurred when crossing the strait 'Pas van Terneuzen' which has a water
depth of 35 m. The tunnel, having an overburden of 15 m in the sands and
clays below the strait, is located with its invert at a depth of more than 60 m
below sea level (figure 9). At the deepest point the tunnel is located in the
deep layer of highly permeable tertiary sands. Never before in the history of
soft ground tunnelling had a tunnel advance with a slurry shield taken place at
such a great depth. This resulted in extraordinary demands on the technology
of the shield machine and on the planning of maintenance operations at the
tunnel face.

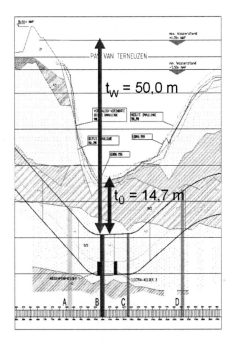

Figure 9: Detail of the longitudinal section of the Westerscheldetunnel at the deepest point

In the planning stage of the project it already became evident, that for maintenance reasons, i.e. to change the cutting knives of the cutterhead, any access
to the excavation chamber would have to take place while maintaining a support pressure of about 6.9 bar. This pressure consists of the highest value of
earth pressure at the tunnel face multiplied with a safety factor of 1.75 plus
the ground water pressure of the regular tidal high water multiplied with a
safety factor of 1.05. When using compressed air at a pressure of 6.9 bar,
the relatively low overburden of only about 15 m would not have permitted
a sufficient safety against blowouts.

The safety against blowout is calculated as the ratio of the vertical stress of the overburdening ground at the crown of the tunnel divided by the support pressure of the compressed air. The safety factor must not be below 1.1, which would have been the case at the deepest point of the Westerschelde tunnel, if compressed air were to be used.

Besides this technical reason, there are also medical and economical reasons which do not permit such a high support pressure with compressed air. In the medical sense, at pressures of more than 3.6 bar the danger of nitrogen narcosis becomes too great. For that reason i.e. German law forbids any compressed air work at pressures exceeding 3.6 bar.

For higher pressures of up to 4.5 bar, special breathing gases, so-called 'Trimix gases' have to be used, which are inhaled in the pressurized environment by using a special helmet. In economical terms, however, it becomes very ineffective to conduct diving operations at working pressures of more than 4.5 bar, because the ratio of the effective working time versus the prescribed decompression time becomes very unfavourable.

For the section of the Westerschelde tunnel where support pressures of 4.5 bar were to be exceeded it became clear that new ways for the performance of maintenance in the excavation chamber had to be found. Conventional techniques, such as a shaft from the surface were far beyond feasibility because of the great water depth of the 'Pas van Terneuzen' with strong tidal currents and with heavy ship traffic.

The solution was to adapt saturation diving technology from the offshore industry by preparing for divers to enter the excavation chamber while it is still filled with pressurized bentonite, which results in a work environment with zero visibility.

To imagine the situation of working next to a tunnel face of 100 m² deep under the sea level in a steel excavation chamber with zero visibility and difficult access while changing cutter tools with a weight of up to 50 kg may create a strong feeling of uneasiness with anyone not involved in professional offshore diving. This situation, however, became real during the construction of the Westerschelde tunnel and it was handled in a very professional manner by the offshore divers, which had been integrated into the tunnelling team from the beginning of the project.

In the planning phase the use of saturation diving works was only intended to be used in an emergency scenario of a breakdown. This situation was planned to be avoided by preventive maintenance while the shield machine was at a depth which would still permit the use of compressed air or trimix gas at the tunnel face. Especially the cutting tools at the cutterhead were all changed before the deepest section was entered although they mostly showed very little signs of wear.

Ironically, at the very deepest point of the tunnel both shield machines showed deformations due to an excessive and unforeseen earth pressure, which required the countermeasure of increasing the overcut of the cutterhead. Therefore, the scraper teeth at the perimeter of the cutterhead had to be replaced by bigger ones.

When performing saturation diving works [5], the divers, working in teams of three, did not decompress to atmospheric pressure at the end of a working shift. Instead they decompressed in the airlock of the shield machine from the working pressure of about 7 bar to a pressure of 4 bar. Using a standardized so-called Natoflange, a transportable shuttle airlock was docked to the shield machine. After reaching a pressure of 4 bar the divers entered the shuttle, which had an autonomous air supply, and were transported out of the tunnel with one of the tunnel trains. In a separate hall at the surface jobsite area a pressurized habitat-chamber was installed including another possibility to dock the transport shuttle. The habitat contained all the necessary amenities needed by the divers during their resting time until their next shift to change the cutting tools in the excavation chamber of the shield machine. During the whole duration of the saturation diving operations medical staff was permanently on site. The dive took a total of up 16 days, which the divers constantly lived "under pressure" including the decompression phase which alone took a total of five days.

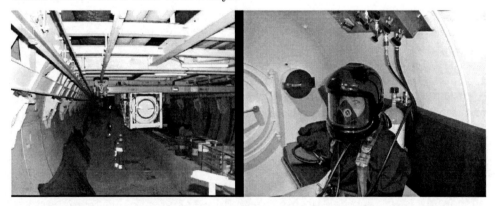

Figure 10: Transport shuttle for saturation dives; diver with trimix gas helmet

This was the first time that a full series of saturation dives had been carried out in a tunnel, and redefined the limits of what can be achieved in modern tunnelling with good planning, advanced equipment design and proper preparation.

5   Mixed tunnel face with soil and hard rock

In this chapter, the experiences of tunnel drives with slurry- or earth pressure balance shields (EPB-shields) in mixed geology, consisting of soft soil and hard rock, is being dealt with. The field of employment of both types of shield machines has been extended considerably in the recent years by various technological improvements. The limits of their applicability are overlapping nowadays and are not exclusively governed by the grain size distribution of the encountered ground anymore [6].

EPB-shields, which use the excavated muck also for the support of the tunnel face, originally were designed for use in fine-grained silty and clayey soils (figure 11). Due to vastly improved conditioning methods of the muck, especially the use of foam as a conditioning agent, EPB-shields today are also used in sands with little content of fines up to sandy gravels [7]. The slurry-shields, which originally were used in coarse-grained and mixed-grained soils, due to improvements in process engineering can nowadays also be used in very sticky clay formations (figure 12; see section 3 and [8, 9]).

Figure 11: EPB-shield for an application in a mixed-face situation with soil and rock

The extension of the application area of both mentioned types of shield machines also includes tunnel drives with a mixed-face situation of soil and hard rock. Tunnel drives with sections of such ground conditions have increasingly been realized in the recent years. Mixed-face conditions with soft soil and hard rock belong to the most challenging environmental conditions for tunnel drives with slurry- or EPB-shields. These conditions are extremely demanding for both the tunnellers and their machine equipment. When being

Figure 12: Slurry-shield for an application in a mixed-face situation with soil and rock

operated under mixed-face conditions both systems show their strengths and deficits very clearly, thereby redefining the limits of their applicability anew.

5.1   Mixed-face conditions for EPB-shields

During regular operation of an EPB-shield in unstable soil the closed mode of operation is chosen (figure 13). In this mode, the excavation chamber is filled completely with the excavated muck and it is kept under pressure according to the necessary support pressure. By conditioning the excavated soil with foam, as it nowadays often is practiced, the muck is turned into a somewhat elastic medium. This elasticity enables the compensation of smaller discrepancies between the inflow of excavated soil into the chamber and the outflow of extracted muck through the screw conveyor, thereby keeping the fluctuation of the support pressure at the tunnel face relatively low.

Using an EPB-shield in a full face situation with rock is a special mode of operation, especially when creating rock chips which cannot be conditioned in the excavation chamber, neither by using conditioning agents nor by the shearing energy exerted by the cutterhead rotation. This special mode of operation sometimes is needed when a desired tunnel trajectory is partly located in unstable soil, which needs active face support, and partly in sections with hard rock. In the transition areas between the soil and the rock then

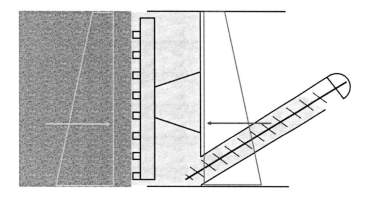

Figure 13: EPB-shield in unstable soil (closed mode)

usually sections of the tunnel occur with a mixed tunnel face. The experience described hereafter has been collected during a tunnel drive for the Deep Tunnel Sewage System in Singapore.

The mode of operation when excavating with an EPB-shield full face in rock is usually the open mode (figure 14) without any support of the tunnel face. In this mode the cutterhead, the steel structure of the excavation chamber and most of all the screw conveyor is exhibited to strong wear because of the rock chips. The amount of wear can be reduced considerably by conditioning with polymer-based conditioning agents (i.e. foam with polymer additive).

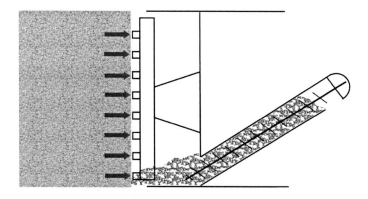

Figure 14: EPB-shield in hard rock (open mode)

The excavation forces on the cutterhead, when cutting full face in rock, are mostly governed by the single contact forces of the disc cutters. With modern

17-inch disc cutters this contact force has a maximum of about 25 tons per disc.

When operating an EPB-shield in mixed face conditions the resulting problems can be distinguished according to the ratio between the amount of soil and the amount of hard rock at the tunnel face. It is assumed that the rock chips excavated from the hard rock section of the tunnel face cannot be plastified by the shearing action of the cutterhead.

When entering a transition zone from soil into hard rock the share of rock at the tunnel face is still low (figure 15). The muck filling of the excavation chamber in this case consists mostly of excavated soil with only some isolated rock chips. The particular problem in this situation is the possible overloading of the peripheral disc cutters (caliber discs), because usually in EPB-mode high advance penetration rates (mm/rev) of the cutterhead are being used.

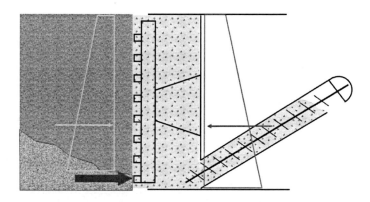

Figure 15: EPB-shield in mixed face with small amount of rock (closed mode)

The reason for the potential overloading of the discs comes from a systematic conflict between the best operation mode for soil and the best operation mode for rock. In closed EPB-mode, a fairly high penetration rate of 20 to 40 mm/rev is desirable, because the lower rotation speed of the cutterhead will minimize wear and the muck is being heated up to a lesser degree due to the lower input of energy. In open hard rock mode the highest possible penetration is governed by the highest possible load on the disc cutters. In sound rock the penetration rate may have to be reduced to values of 10 mm/rev or lower. Is therefore a rising rock bed being approached by an EPB-shield in closed mode with high penetrations, extreme overloading of the peripheral disc cutters may occur resulting in heavy damage of the discs.

The first contact of the peripheral discs of a cutterhead with a rising rock bed is very difficult to detect from the online machine data. The resulting difference of the cutterhead thrust from the increased loads onto some peripheral cutters is almost impossible to detect because of the overlaying and fluctuating earth pressure, which results from the closed mode of operation and which can be much higher than the additional contact forces with the rock.

Is the percentage of the rock at a tunnel face larger than 50%, the problem of overloading single disc cutters decreases in importance, since already a large number of discs is in contact with the rock part of the tunnel face.

In a situation where the rock covers the largest part of the tunnel face (figure 16) a switch to open mode is desirable. This mode, however, can only be chosen, when the remaining part of soil at the tunnel face is stable without any support and if the inflow of ground water is very low.

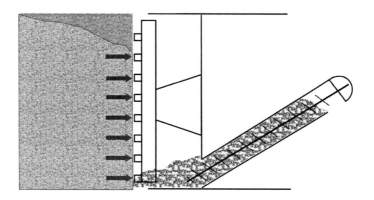

Figure 16: EPB-shield in mixed face geology with high amount of rock (open mode)

The most problematic situation will occur when only a small part of soil is remaining at the tunnel face, which is unstable and needs active face support, and when there is a significant inflow of water because of the soil being highly permeable (figure 17). In this situation it has to be attempted to create an earth-pressure support of the tunnel face with excavated spoil, which mainly consists of rock chips and only a small amount of fines [10], resulting in extremely high wear.

Such a closed mode of operation in predominantly hard rock conditions requires, as mentioned before, a comparatively high revolution speed of the cutterhead, which results in a conversion of a very high amount of kinetic energy into heat energy in the excavation chamber. The resulting heating of the muck and the excavation chamber to temperatures that may well be in the range from 70 to 90°C making time-consuming cooling phases by extra ventilation necessary before the excavation chamber can be entered by

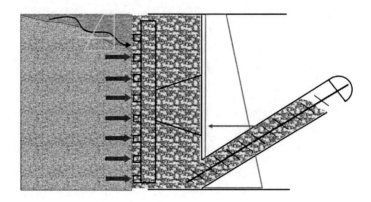

Figure 17: EPB-shield in mixed face geology with high amount of rock (closed mode)

humans for necessary maintenance works. On the other hand maintenance work, especially cutter inspections and cutter changes, become necessary far more frequently under these adverse ground conditions. This may result in a drastic decrease of the daily excavation performance during advances in sections with mixed-face geology.

In case the rock chips have sufficient strength so that they cannot be ground to fines, large pore spaces will be present between the rock chips, making it very difficult and maybe even impossible to stabilize the muck against the inflow of ground water into the excavation chamber, even when using all available means of conditioning [11]. Also, the necessary pressure gradient along the screw conveyor, which is needed to build up an active face support pressure, cannot be created in this situation. This can result into an uncontrolled inflow of groundwater into the excavation chamber, increasing the chance of erosion and instability at the remaining part of the tunnel face consisting of soil. Even by accepting heavy secondary wear, caused by a closed mode in mixed face situations, it sometimes has been impossible to stabilize the remaining soil fraction [12].

Also maintenance work for tool changes under compressed air can be very difficult in mixed face situations. One reason may be a high loss of compressed air, when the soil part of the tunnel face is very coarse or the rock part is highly fractured. In this case, additional measures, such as a treatment of the tunnel face with a cement-bentonite-mixture (figure 18) may become necessary in order to decrease the permeability of the ground against compressed air.

Finally, driving a tunnel in mixed face geology with an EPB-shield may also result in difficulties with the grouting of the annular gap behind the tailskin. When the cutterhead is entering the rock and no active support pressure

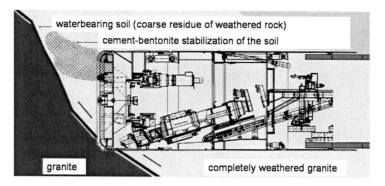

Figure 18: EPB-shield in mixed face geology with high water ingress and cement-bentonite injection

is maintained in the excavation chamber anymore while the tailskin is still situated in soft soil, requiring active grouting, the grout may start to flow along the skin of the shield into the excavation chamber because of the missing counter pressure. As a consequence, additional water ingress may occur and the segment ring may not be fully embedded in grout anymore, resulting in higher deformations of the segment ring and also in higher deformations at the surface (figure 19).

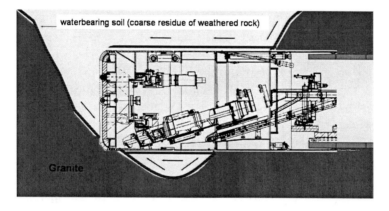

Figure 19: EPB-shield in mixed face geology with water ingress from the tail skin

5.2 Mixed-face conditions for slurry-shields

Tunnel drives with slurry shields in sections with mixed-face geology in many cases are easier to manage. The face support with a fluid, mostly a bentonite suspension, in most cases permits full control over the stability of the tunnel face (figure 20). Also maintenance access to the excavation chamber in most cases will be possible because of the sealing effect of the bentonite suspension on a permeable tunnel face. Additional costly and time-consuming efforts to avoid the loss of compressed air at the tunnel face usually are not necessary.

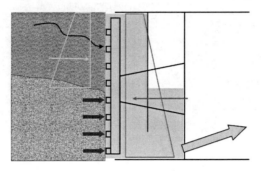

Figure 20: Slurry shield in mixed face geology

A risk that also remains for slurry shields is the possibility of overloading single disc cutters when entering a transition zone from soil into rock. The systematic conflict regarding the revolution speed of the cutterhead and the resulting penetration rate, however, is smaller for a slurry shield in comparison with an EPB-shield. The filling of the excavation chamber with bentonite suspension permits a fairly free adjustment of the cutterhead revolution speed in correspondence with the desired penetration rate.

A successful example for a slurry shield drive with long sections under mixed face conditions is the Airport Rail Link project in Sydney, Australia (see figure 12). Here the cutterhead besides soft ground knives was equipped with a full set of disc cutters for the hard rock sections of the project. When designing a slurry shield for such ground conditions, special attention has to be paid to the suction inlet area in order to avoid blocking by larger rock chips, which in case of the mentioned project were likely to be combined with chunks of clay. Here the combination of a rotary stone crusher and agitators has proven to be ideal (see figure 7). In the case of the Sydney project the tunnel has been partly been built under the runways of the international airport without any damages while the operation of the airport remained unaffected.

5.3   Improvements for shield drives in mixed-face geology

The danger of creating an excessive overload and destroying the disc cutters when partially cutting into a rock bed (figure 21) can be reduced by driving the shield machine in a penetration-controlled manner when approaching the rock. This, of course, will only be possible when the geological exploration gives a sufficient amount of data to predict the transition zones between soil and rock with a reasonable accuracy.

Figure 21: Damaged disc cutter

The highest permissible penetration when approaching the rock can be calculated based on the correlation established by Rostami [13] between the rock strength and the penetration of a disc cutter. This correlation also considers the geometry of the cutter ring and the contact force of the cutter. From this correlation a graph was developed (figure 22), which indicates the maximum penetration of a disc cutter in dependency of the rock strength if an overloading of the cutter has to be avoided.

In reaction to frequently encountered difficulties with the grouting of the annular gap during EPB-shield drives in mixed face geology an additional possibility for secondary grouting has proven to be very helpful. In this case the segments have to be equipped with special grout openings including a simple valve to avoid the backflow of grout.

An additional measure to improve the ability to support the tunnel face and to control the inflow of groundwater in sections with a high proportion of rock at the face might be the use of an extra-thick paste-like suspension, which is injected into the excavation chamber. This measure, however, would require that 50 to 100% of the excavated ground volume has to be injected in addition. This would create an enormous additional logistical and technological effort.

Another possibility to improve the suitability of EPB-shields for a use in mixed face geology would be the additional use of a pressurized slurry transport

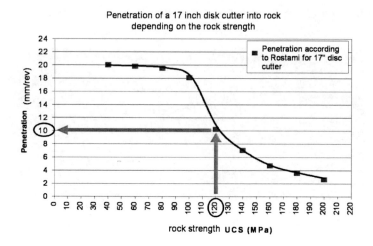

Figure 22: Correlation between rock strength and penetration for a 17 inch disc cutter at limit load

circuit, which would be connected to the outlet of the screw conveyor in a closed system. The muck would then pass a stone crusher after exiting the screw conveyor before it enters a box, which is connected to a slurry circuit similar to the circuit used with slurry-shields. It remains questionable and should be the topic of intense individual evaluations whether the effort, which would be necessary to install a slurry circuit and a the necessary separation plant, would not justify the use of a slurry shield in the first place.

5.4   Comparison of the two systems

If a tunnel drive has to be performed in a geology which will include a mixed face situation combining soft soil and hard abrasive rock, the use of a slurry shield will be favorable regarding the following aspects:

- stability of the tunnel face

- reduction of surface settlements

- avoiding the uncontrolled ingress of ground water into the excavation chamber

- reducing secondary wear at the cutting tools, the cutterhead and the excavation chamber

- enabling access to the excavation chamber for maintenance purposes

- time consumption for preparing and performing maintenance in the excavation chamber

- ensuring a complete grouting process of the annular gap.

A tunnel drive with an EPB-shield in comparison has the following positive aspects:

- requires a lower capital investment into the technical equipment, especially by working without a slurry circuit and a separation plant

- requires less operation cost for running and maintaining the tunnelling equipment

- requires a smaller number of highly skilled tunnelling personnel due to an overall less sophisticated level of the tunnelling equipment, which may be an important factor in less industrially developed regions of the world

- disposal of the muck in many cases is less costly.

While the advantages of the slurry shield mostly are of a technical and risk-limiting nature the advantages of an EPB-shield predominantly are on the economical side. The economical advantages, however, will only persist throughout a project, if the risks of using an EPB-shield in mixed face geology will not result in major unforeseen damages or delays.

Against the background of international experiences of the recent years [12], showing that in some cases the risks of using an EPB-shield has been underestimated, the following procedure for the planning of tunnel drives, which include mixed face sections, is recommended: Primarily, the use of a slurry shield should be considered. The alternative use of an EPB-shield should only be considered after performing a thorough risk analysis and a very detailed evaluation of the lower cost of the EPB-system in comparison with the necessary cost for adequate provisions to cover the potentially higher level of risk.

Accordingly, for tunnel drives including mixed face sections the following rule of thumb applies: The less information is available about the precise geological situation, and the lower the possible degree of detail for a corresponding risk analysis, the more advisable is the use of a slurry shield.

The mentioned risk analysis and comparative cost analysis should be performed by the client and his experts before the beginning of a tendering process, and should lead to a definitive requirement for the tunnelling technology to be applied. Otherwise, it will be possible that also in the future, due to the cost pressure in the tunnelling industry, a type of shield machine will be applied where the limits of its applicability will be exceeded to the damage of the whole project.

6   Joint code of practice for risk management of tunnel works

Tunnelling is a profession followed enthusiastically by most of those who work in it. This at least to a certain extent can be attributed to the fact, that in tunnelling very challenging risks have to be analyzed and managed while working on large scale infrastructure projects. Especially when tunnelling in an urban environment enormous damage may occur for all parties involved if not all of the associated significant risks were identified beforehand and suitable measures were taken to prevent damage.

In October of 2001 the Association of British Insurers (ABI) contacted the British Tunnelling Society (BTS) regarding their growing concerns about recent losses associated with tunnelling works. The insurers had the impression that the tunnelling industry had deficits at risk management. A fundamental improvement of risk management principles was required so that insurance could be provided for tunnel works in the future.

Instead of withdrawing completely from providing insurance cover for tunnelling projects, or increasing insurance fees or restricting insurance cover, the insurers developed in cooperation with the British tunnelling industry a 'Joint Code of Practice' for better management of the risks related to tunnelling activities.

The goal of the code is to promote best practice for the minimization and management of risks which are related to the design and the construction of underground structures, such as tunnels, stations, shafts, and caverns.

The fundamental principles, which are defined by the Code, include:

- Identifications of hazards and associated risks are carried out on a project-specific basis during each stage of a project.

- Risk assessments are recorded in risk registers. At each stage of a project the responsible party for the control and management of each identified risk is defined.

- Identified risks are managed to ensure their reduction to a level as low as reasonably practicable.

- Risk registers are continuously reviewed and updated. It is ensured that the involved parties are familiar with the identified hazards and risks at any time during a project.

The 'Joint Code of Practice for Risk Management of Tunnel Works in the UK' [1] was published in 2003 and can be ordered under www.britishtunnelling.org. The code has been approved and accepted by the Association of British Insurers for application at future tunnel works in the UK. Currently, an international version of the code is being prepared in cooperation with the International Tunnelling Association (ITA).

References

[1] British Tunnelling Society: Joint Code of Practice for Risk Management of Tunnel Works in the UK. British Tunnelling Society, 2003, www.britishtunnelling.org

[2] Thewes, M.: Adhäsion von Tonböden beim Tunnelvortrieb mit Flüssigkeitsschilden, Berichte aus Bodenmechanik und Grundbau Vol. 21, Bergische Universität Wuppertal, FB Bauingenieurwesen, 1999 Shaker Verlag, Aachen, ISBN 3-82565-6402-2

[3] Herrenknecht, M.: Praktische Hinweise zur Anpassung von Tunnelvortriebsmaschinen an die Erfordernisse beim Vortrieb; Forschung + Praxis 39, Proceedings of the STUVA-Symposium 2001 in Munich, 2001, Bertelsmann-Springer Bauverlag, Gütersloh

[4] Sager, H.-J.; Maidl, U.: Innovative Methoden zur Begegnung der Verklebungsproblematik bei den Flüssigkeitsschilden der Tunnel unter der Westerschelde; Forschung + Praxis 38, Proceedings of the STUVA-Symposium 1999 in Frankfurt a.M., 2000, Alba Verlag, Düsseldorf

[5] Le Pechon, J., Sterk, W., van Rees Vellinga, T.: Saturation diving for tunnelling operations. Proceedings AITES/ITA World Tunnel Congress, Milano, June 10-13th 2001,

[6] Maidl, U.: Beitrag zur Abgrenzung der Einsatzbereiche für Erd- und Flüssigkeitsschilde. *Bauingenieur 76*, März 2001, pp. 118-122

[7] EFNARC: Specification and Guidelines for the use of specialist products for Soft Ground Tunnelling. 2003, EFNARC, Association House, Farnham, Surrey, UK

[8] Thewes, M.: Adhäsion von Tonböden beim Tunnelvortrieb mit Flüssigkeitsschilden. *Geotechnik 04/2003*

[9] Thewes, M.; Burger, W.: Verklebungen beim Schildvortrieb in Tonformationen - Erkennen und Begrenzen technischer und vertraglicher Risiken. Forschung + Praxis 40, Vorträge der STUVA-Tagung 2003 in Dortmund, Bertelsmann, Gütersloh, 2003

[10] Della Valle, N.: Boring through a rock-soil interface in Singapore. 2001 RETC Proceedings, pp. 633-645

[11] Herrenknecht, M.; Liebler, B.; Maidl, U.: Geotechnische und mechanische Interaktion beim Einsatz von Erddruckschilden im Fels. Forschung + Praxis Nr. 40, Tagungsband der STUVA-Tagung 2003, Dortmund, pp. 175-181

[12] McFeat-Smith, I.: TBM selection for control of water ingress and face stability for tunnelling in the widest range of geological conditions. Proceedings of Underground Singapore 2001, Singapore, 29-30. Nov. 2001, pp. 159-168

[13] Rostami, J.; Gertsch, G.&L.: Rock fragmentation by disc cutter: a critical review and an update. NARMS-TAC 2002, Hammah et. al. (eds.), University of Toronto, 2002, pp. 977-985

Earthquake resistant design of tunnels

Christos Vrettos

Technical University of Kaiserslautern, Division of Soil Mechanics and Foundation Engineering, 67663 Kaiserslautern, Germany

Abstract: Although tunnels are inherently less sensitive to seismic effects than surface structures, a rational and consistent design methodology is required to guarantee the high standards posed on major infrastructures. The effects of seismic motion on tunnels in terms of faulting, shaking and ground failure are presented and appropriate countermeasures are briefly described. Design strategies, ground motions and geotechnical properties are discussed. The deformation method is elucidated and the importance of soil-structure interaction is demonstrated. Simple state-of-the-art methods for analyzing the tunnel response to ground shaking are described for axial and curvature as well as for ovaling/racking deformations.

1 Introduction

In general, tunnels are considered less vulnerable to seismic actions than aboveground structures, like high-rise buildings or bridges. Due to the embedment, the surrounding ground inherently provides protection against earthquake damage by restricting the free movement of the tunnel structure. The good performance of tunnels was demonstrated during the 1986 Mexico City earthquake, where subway structures in soft soils remained undamaged in contrast to other surface facilities that suffered severe damage. Nonetheless, tunnels are special structures not covered by the seismic building codes. Ground properties usually vary along the alignment, and the fact that due to the effects of spatial variability of earthquake motion the tunnel response analysis requires the consideration of a structural 3D global tunnel model calls for a more rational design approach to earthquake hazards. While deep level tunnels in rock are considered safe, special attention is required for shallow tunnels in soft soils, like for example immersed tubes crossing shallow waters. A rough estimate of the damage sustained by tunnel structures in seismically active areas may be obtained from the results of the study carried out by Power *et al.* [13]. They related the measured surface peak ground acceleration to the observed damage identifying that damage was slight up to a value of 0.2 g.

The inherent problem in such case studies is the difficulty to isolate in the back analysis the loads due to earthquake effects from the usual static loads. An alternative offers the instrumentation of the tunnel structure and the direct measurement of the seismic action, that is however costly and thus restricted to prestigious projects. Underground subway stations were severely damaged during the 1995 Hyogoken-Nanbu earthquake in Kobe. The most spectacular among them is the complete collapse of the Daikai Subway Station. Nonlinear analyses of the failure mechanism concluded that the reason was the collapse of the center column by the combination of bending and shear.

The subject of this paper is to provide a summary of the earthquake effects on tunnels, of the design principles, and of the state-of-the-art methods to analyse the seismic tunnel response.

2   Effects of ground motion on tunnels

The effects of earthquakes on tunnel structures may be categorized into three groups: faulting, ground failure, and shaking.

Faulting: It refers to direct primary shearing displacements of bedrock that is usually restricted to relatively narrow seismically active fault zones. Sliding along a geologic fault produces stresses that may be significantly higher than the magnitude induced by shaking. In general it is not practical to design a tunnel to restrain major displacement in the order of several centimetres or meters. Tunnel alignements should avoid active faults whenever possible. If not, one has to accept the induced displacement, localize the damage and provide means to facilitate future repair works in order to avoid unnecessary long shut-down of the system after a major earthquake.

Ground failure: Damage associated with ground failure may be caused by liquefaction of soils, slides of rock or soil, soil subsidence, and other effects of ground motion. These hazards may be reduced by careful site investigations along the tunnel alignment. The major threat refers to liquefaction that is however restricted to shallow soil depths. Methods to estimate the liquefaction potential are nowadays well-established and described in the relevant codes. Typical example is the site of San Francisco's MUNI metro underground extension which was reclaimed by filling over the seabed in the late 1800's, cf. Arango *et al.* [2]. The strongly inhomogeneous site experienced twice liquefaction in 1906 and also the more recent 1989 Loma Prieta earthquakes. Lateral spreading reached values of one meter and half a meter, respectively. Excessive ground improvement works were necessary to minimize liquefaction hazard and its impact on the tunnel structures. Portals are particularly vulnerable to permanent displacements. Portal slopes can be reinforced e.g. using tiedback anchors. Large blocks of rock loosening may be secured individually or by applying shotcrete.

Shaking: Damage due to shaking for lined tunnels may include spalling, crack-ing or failure of the liner. Shaking may also lead to a reduction of shear strength of the soil or rock mass above the tunnel. In this case the tunnel support system has to withstand additional loads. For unlined tunnels earth-quake motion may cause block motion, spalling, rock fall, or local opening of joints. The seismic response of such underground structures depends on the shape and depth of excavation, the mechanical properties of the surrounding soil and rock mass, and the intensity and characteristics of seismic action. Dowding and Rozen [5] suggest that tunnels in rock are not expected to ex-perience damage as long as the particle velocity due to ground motion is less than 20 cm/sec. Due to the strong variability of natural soils and the inher-ent difficulty to accurately predict the dynamic soil behaviour, such threshold values can not be defined for sedimentary soil deposits.

Since damage due to ground failure can be mitigated by appropriate ground improvement methods, engineering design mainly addresses protection against fault rupture and excessive ground motion.

Countermeasures against potential offsets due to faulting typically consist of excavating an oversize section through the fault zone. These oversized fault chambers shall be sufficient to realign the track/road with acceptable lateral and vertical curves subsequent to a major fault rupture event, while reinforc-ing the ground in and around the shear zone in order to prevent collapse. A ground reinforcement system of great ductility, such as a combination of lattice girders, wire mesh, rock dowels, and shotcrete is favourable. Opera-tional considerations should include instrumentation of the fault crossings to detect fault creep or offsets. The amount of likely slip shall be determined by a seismic hazard study. For preliminary design purposes rough estimates based on the relations by Wells and Coppersmith [16] may be used.

Loads due to active fault displacement are not amenable to calculation. Hence, ground shaking is the focus of the following sections.

Shaking during an earthquake is transmitted from the bedrock to the ground surface primarily by shear waves, the orientations of which are somewhat ran-dom with respect to the structure considered. During their passage through the ground these waves interact with each other yielding complicate wave deformation patterns before impinging upon an underground structure at various angles. Shear waves propagating parallel to the longitudinal axis of a long tunnel structure cause transverse displacements producing bending dis-tortion in the structure. Such waves travelling at right angles to the structure will move it back and forth longitudinally, and may tend to pull it loose at zones of abrupt transitions in soil conditions where wave properties suddenly change. Diagonally impinging waves subject different parts of a long structure to out-of-phase components of displacement in both its longitudinal and transverse directions.

3   Definition of seismic action

The seismic input motion is expressed in terms of ground accelerations, velocities, displacements, and response spectra. This information is obtained on the basis of a project-specific seismic hazard analysis. The seismic input is generated for the rock-outcrop conditions. The seismic hazard analysis procedure consists of several steps that typically consider: i) the contributing seismic sources in the near region of the site with the associated upper-bound magnitudes, ii) the attenuation relationships for acceleration response spectra, iii) magnitude-recurrence relation for each source zone, iv) fault-rupture-length vs. magnitude relation. Response spectra are determined for various return periods, or, equivalently, probabilities of exceedance in a specific period of time.

Next, a free-field seismic site response analysis is carried out under the assumption of vertically propagating horizontally polarized shear waves. It is a 1D-analysis with only one degree of freedom, i.e. the horizontal displacement. A popular code for this computation is SHAKE that uses a visco-elastic soil model and an equivalent linear iterative procedure to represent the nonlinear variation of shear modulus and damping with shear strain level. The seismic input is applied in terms of an acceleration time history at a hypothetical rock outcrop. Results obtained are soil displacements, shear stresses and strains, as well as strain compatible shear moduli and damping ratios.

As will be evident later in this text the relevant kinematic quantity for the design is the ground displacement expected to occur during the earthquake, whereas the input to SHAKE is defined in terms of acceleration. Thus, particular attention shall be paid in selecting an earthquake record that realistically captures displacement, velocity, and acceleration to be used in the analysis.

4   Geotechnical exploration and ground parameters

The geological setting is usually determined during the initial design phase of the project. Standard geotechnical and geophysical investigations to be carried out are identical to those usually performed in non-seismically active areas. Essential for the seismic design of tunnels in sedimentary soils is the depth of bedrock, whereby one has to distinguish between hard rock, encountered in large depths, and soft rock found in shallower depths. Dynamic stiffness of the ground is mainly represented by the shear wave velocity c_S with values ranging from over 800 m/s for sound, non-weathered non-fractured rock to 150 m/s for soft cohesive soils and even lower for soft organic soils.

The compressional wave velocity c_P is also used in practice, mainly for rocks, since the value of this parameter in soft soils is dominated by the presence of ground water in the pores, thus not enabling a refined material characterisation. A widely accepted classification system covering the entire range from rock to soil still does not exist. A proposal is found in the AFPS/AFTES Guidelines [1] that attempts also a link to parameters describing rock fracturing. An extract is given in Table 1.

Ground type	q_c (MPa)	N_{SPT} (-)	E_M (MPa)	U_c (MPa)	c_S (m/s)	c_P (m/s) below GWT	c_P (m/s) above GWT	RQD (%)	ID (cm)
Sound, unweathered, unfractured rock	-	-	-	>10	800		>2500	>75	>60
Weathered or fractured rock	-	-	50-100	6-10	500-800		1000-2500	50-75	20-60
Cohesive soil (stiff clay or marl)	>5	-	>25	>0.4	>400		>1800	-	-
Compact granular soil	>15	>30	>20	-	>400	>1800	>800	-	-
Decomposed or severely fractured rock	-	-	50-100	1-6	300-500	-	400-1000	<50	<20
Moderately compact granular soil	5-15	10-30	6-20	-	150-400	1500-1800	500-800	-	-
Moderately consistent cohesive soil and very soft rock	1.5-5	-	5-25	0.1-0.4	150-400	-	1000-1800	-	-
Loose granular soil	<5	<10	<6		<150	<1500	<500	-	-
Soft cohesive soil (soft clay or mud)	<1.5	<2	<5	<0.1	<150	<1500	<500		

Table 1: Ground classification from AFPS/AFTES Guidelines (extract)

In Table 1 q_c is the tip resistance of the CPT test, N_{SPT} the blow count of the SPT test, E_M is the Menard pressuremeter modulus, U_c is the unconfined compressive strength, RQD is the Rock Quality Designation, and ID is the interval between discontinuities (joint spacing). Deformability of sound rock is recommended to be determined from dilatometer tests. As can be seen, the range of values is large, thus demonstrating the need of project-specific tests in order to obtain more reliable values of these parameters. Density ρ can easily be determined by standard methods.

Seismic site response is usually calculated using an equivalent linear analysis that requires knowledge of the variation of dynamic soil stiffness with strain level. The dynamic soil properties entering the calculation are the shear modulus G and Poisson's ratio ν. They can be determined from c_S and c_P, but in most cases this yields unrealistic values for the Poisson's ratio.

$$c_S = \sqrt{\frac{G}{\rho}} \quad ; \quad c_P = \sqrt{\frac{2(1-\nu)}{1-2\nu}\frac{G}{\rho}} \tag{1}$$

The alternative, that is usually adopted, consists in first determining G from c_S and density ρ and then assuming a realistic value of ν from own experience and published recommendations.

The amplification of the seismic bedrock motion within a soil sediment, and accordingly the displacement imposed on the tunnel structure, is strongly influenced by the impedance ratio $\alpha_{s/r}$ defined by

$$\alpha_{s/r} = \frac{\rho_s c_{S,s}}{\rho_r c_{S,r}} \tag{2}$$

where $c_{S,s}$, ρ_s and $c_{S,r}$, ρ_r are the shear wave velocities and densities of the soil and the underlying rock, respectively. Amplification increases as the impedance ratio becomes smaller, cf. Kramer [8]. Equation (2) assumes vertically propagating shear waves.

For tunnels embedded in soft soils the geotechnical exploration program is enriched by suitable dynamic tests. In-situ down-hole or cross-hole tests are executed to determine wave velocities of the various layers at small strain levels. Seismic cone penetration tests, i.e. a combination of CPT and down-hole tests become increasingly popular. Less expensive are geophysical measurements on the ground surface such as seismic refraction surveys. The measured values shall be compared with estimates obtained from empirical correlations between shear wave velocity and cone penetration resistance or SPT blow count numbers, as given by Kramer [8].

The nonlinearity of the dynamic soil parameters, i.e. the variation of shear modulus and damping ratio with shear strain amplitude, can be obtained from laboratory resonant column tests. An accurate soil modelling should account for the effects of plasticity index and effective confining pressure. Cyclic simple shear or triaxial tests are performed in order to assess the soil degradation due to repeated loading and the liquefaction potential in water saturated loose sands and silty sands. Testing procedures as well as some design equations and curves are summarized by Kramer [8].

Usually shear wave velocities measured during in-situ seismic tests will be larger than those determined in the laboratory. A method often used in practice is to adjust the small strain values to the in-situ values and use the shear-strain dependency as determined in the laboratory. Finally, it should be kept in mind that dynamic tests, in particular those conducted in the laboratory, are expensive, so that judgement is required in selecting representative samples for the testing program.

5 Design principles

In general, earthquake resistant design is carried out for two levels of seismic motion.

Level 1 is characterized by an earthquake of moderate intensity that is assumed to occur several times during the structure's lifetime. For this level it is required that the structure behaves more or less elastically and suffers little or no damage during the earthquake event.

Level 2 represents a major earthquake with small probability of exceedance during the life of the structure (e.g. 5%). For this level the structure shall designed to avoid collapse, but structural degradation due to post-elastic deformations is allowed.

The requirements for the two design levels are described by the relevant national codes, or in the design specification of the project. Each design level is associated with a particular loading combination, whereby the load factors depend on the desired performance of the structural members. Typical load combinations are summarized by Hashash et al. [7].

Due to this dual design strategy the relevant design criteria may be different for medium and high seismicity regions, respectively. In medium seismicity regions the difference in response spectra between level 1 and level 2 hazard is large and level 2 will control the structural design. In high seismicity regions, on the other hand, this difference becomes small and level 1 may govern the design. Thus, both design levels shall be considered in order to ensure that the specified dual performance criteria are met.

Unlike the case of surface structures whose seismic response is governed by inertia effects, the response of tunnels embedded in the ground is primarily kinematic, i.e. it is caused by the compatibility of the tunnel deformation to that of the surrounding ground. Therefore, soil-structure interaction effects are essential to the design.

In static design the loads are well-known and the analysis is carried out in terms of forces (force method). In contrast, seismic structural response due to the imposed seismic displacements (deformation method) strongly depends on structural details. Seismic forces acting on structural members increase as the structure's flexibility decreases. Therefore the lining is designed for maximum flexibility that is achieved by improving ductility using adequate materials and reinforcement. Plastic hinges are allowed at critical joints after checking that no plastic hinge combination leads to a collapse mechanism.

Design is carried out for three principal types of deformation as depicted in Figure 1:

- Axial extension and compression due to wave motions parallel to the tunnel axis causing alternating compression and tension.

- Longitudinal bending due to curvature imposed by those components of the seismic wave that produce particle motions perpendicular to the tunnel axis.

- Ovaling for circular, and racking for rectangular tunnels, respectively, that develops when shear waves propagate normal or nearly normal to the tunnel axis, resulting in a distortion of the cross-sectional shape of the tunnel lining.

6   Axial and curvature deformation analysis

We first consider axial and curvature deformations due to this snaking mode in the direction along the tunnel axis: The structure moves like a snake subjected to the earthquake ground wave motion. Essential to the analysis is the relative stiffness of a tunnel structure and the surrounding soil. This relative stiffness of a circular lining has been described by means of two separate and distinct ratios. The compressibility ratio C relates the compressional/extensional stiffness of the soil medium relative to that of the lining, while the flexibility ratio F is a measure of the flexural stiffness (resistance

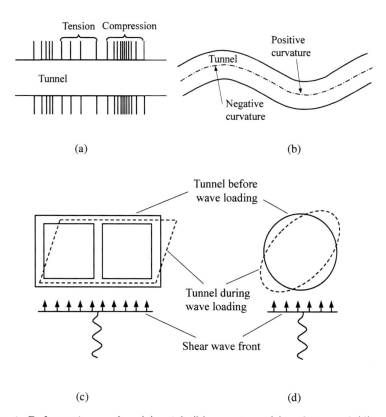

Figure 1: Deformation modes: (a) axial, (b) curvature, (c) racking, and (d) ovaling

to ovaling).

$$C = \frac{E(1 - \nu_l^2)d}{2E_l t_l(1 + \nu)(1 - 2\nu)} \quad ; \quad F = \frac{E(1 - \nu_l^2)d^3}{48E_l I_l(1 + \nu)} \tag{3}$$

where E and ν are the modulus of elasticity and the Poisson's ratio of the ground, E_l and ν_l the respective values for the tunnel lining, I_l is the moment of inertia of the lining, and t_l and d the thickness and the diameter (or equivalent diameter) of the circular lining tunnel, respectively.

It has been shown that compressibility has little effect on the structural tunnel behaviour and that the flexibility ratio is large enough ($F > 20$) to justify the assumption that – except for reinforced concrete lining in soft soils – the liner offers little or no resistance to ground motion, i.e. the liner conforms to the ground motion. Imposing the free field strain on the structure is therefore a conservative assumption. However, for stiff tunnel structures in soft soils, ignoring the effects due to difference in flexibility leads to large sectional forces and non-economical design calling for proper consideration of soil-structure interaction. We first consider structures that conform to ground motion.

The behaviour of the lining may be simulated as a buried structure subject to ground deformations under two-dimensional plane-strain conditions. The analysis follows the methodology proposed by Newmark [11] and Kuesel [9].

6.1 Structure conforming to ground motion

We assume a sinusoidal motion of the ground with wave length λ and amplitude D_0. The wave has an angle of incidence ψ with respect to the tunnel axis, cf. Figure 2. The wave can be either a body shear wave or a Rayleigh surface wave. Compressional waves are usually considered not critical for the design.

The displacements parallel (x) and perpendicular (y) to the tunnel axis are:

$$u_x = D_0 \sin \psi \sin \left(2\pi \frac{x}{\lambda} \cos \psi - \omega t \right) \tag{4}$$

and

$$u_y = D_0 \cos \psi \sin \left(2\pi \frac{x}{\lambda} \cos \psi - \omega t \right) \tag{5}$$

with t denoting time and ω the circular frequency.

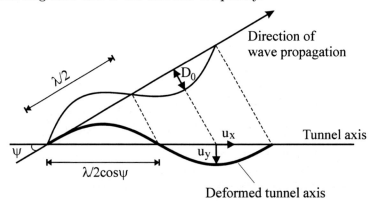

Figure 2: Wave oblique to tunnel axis

The axial distortion of the ground (and accordingly also of the tunnel structure) due to the seismic motion in the direction of the tunnel axis is:

$$\varepsilon_x = \frac{\partial u_x}{\partial x} = \frac{2\pi}{\lambda} D_0 \sin \psi \cos \psi \cos \left(2\pi \frac{x}{\lambda} \cos \psi - \omega t \right) \tag{6}$$

with amplitude:

$$|\varepsilon_x| = \frac{2\pi}{\lambda} D_0 \sin \psi \cos \psi \tag{7}$$

Due to the seismic motion perpendicular to the tunnel axis, the tunnel axis is deformed with the curvature:

$$\frac{1}{R} = \frac{\partial^2 u_y}{\partial x^2} = -\left(\frac{2\pi \cos \psi}{\lambda}\right)^2 u_y \tag{8}$$

Due to this curvature the tunnel lining is subjected to bending that yields at two opposite sides of the tunnel with width b the following compressive and tensile strains, respectively:

$$\varepsilon_b = \pm \frac{b}{2} \left(\frac{1}{R}\right) \tag{9}$$

with amplitude:

$$|\varepsilon_b| = \pm \frac{b}{2} \left(\frac{1}{R}\right) = \frac{b}{2} \left(\frac{2\pi \cos \psi}{\lambda}\right)^2 D_0 \cos \psi = \frac{2\pi^2 D_0 b \cos^3 \psi}{\lambda^2} \tag{10}$$

Thus, the maximum axial distortion of the tunnel lining due to stretching and bending is:

$$\varepsilon = |\varepsilon_x| + |\varepsilon_b| = \left(\frac{2\pi D_0}{\lambda}\right) \left(\sin \psi + \frac{\pi b}{\lambda} \cos^2 \psi\right) \cos \psi \tag{11}$$

It can be seen that the shorter the wave length the larger the induced seismic strains.

Assuming that for wave lengths shorter than six times the tunnel width the behaviour of the tunnel corresponds to that of a rigid body, we obtain for $\lambda / \cos \psi = 6b$

$$\varepsilon = \left(\frac{2\pi D_0}{\lambda}\right) \left(\sin \psi + \frac{\pi}{6} \cos \psi\right) \cos \psi \tag{12}$$

This function has its maximum for $\psi = 32°$. The corresponding strain value is

$$\varepsilon_{\max} = 5.2 \frac{D_0}{\lambda} \tag{13}$$

Expressing the displacement amplitude D_0 in terms of the particle velocity amplitude V_0 and the particle acceleration amplitude A_0

$$D_0 = \frac{V_0}{c} \frac{\lambda}{2\pi} \quad ; \quad D_0 = \frac{A_0}{c^2} \left(\frac{\lambda}{2\pi}\right)^2 \tag{14}$$

where c denotes the effective (apparent) propagation velocity of the considered wave, it can be seen that the axial strain ε_x is governed by the particle velocity, while curvature and accordingly bending strain ε_b are governed by the particle acceleration. Since both strain components contribute to the resulting sectional forces, it is essential to select a seismic input motion with appropriate realistic values both for V_0 and for A_0. This requires careful selection of the seismic input motion used in the seismic site response analysis to calculate the free-field response.

A point that deserves attention refers to the value of the effective wave propagation velocity in the horizontal direction, as used e.g. in equations (14). It is not necessarily the same as the wave velocity through the soil overburden. Observations indicate values in the km/s range. The reason for these high values lies in the fact, that depending on the thickness of the soil stratum the effective wave velocity of the near surface ground is affected by the much higher wave velocities of the deep-lying hard strata.

Sectional forces in the tunnel structures are derived by beam theory. The peak values are as follows:

- Axial force (compression-tension):

$$|N| = \frac{2\pi}{\lambda} \sin \psi \cos \psi E_l A_l D_0 \tag{15}$$

- Bending moment:

$$|M| = \left(\frac{2\pi}{\lambda}\right)^2 \cos^3 \psi E_l I_l D_0 \tag{16}$$

- Shear force:

$$|Q| = \frac{2\pi \cos \psi}{\lambda}|M| \tag{17}$$

In equation (15) A_l is the cross-sectional area of the tunnel lining.

The maximum axial force is obtained for $\psi = 45°$ while maximum bending moment and shear force occur for $\psi = 0°$. The corresponding values are:

$$|N|_{\text{max}} = \frac{\pi}{\lambda} E_l A_l D_0 \tag{18}$$

$$|M|_{\text{max}} = \left(\frac{2\pi}{\lambda}\right)^2 E_l I_l D_0 \tag{19}$$

$$|Q|_{\text{max}} = \frac{2\pi}{\lambda}|M|_{\text{max}} \tag{20}$$

6.2   Consideration of soil-structure interaction effects

In many cases the tunnel structure is stiffer than the surrounding soil medium. Accordingly, it will distort less than the free-field ground deformation and there will be an interaction between the tunnel structure and the surrounding medium. To model these effects we assume that the tunnel structure behaves like an elastic beam supported by series of idealized springs acting in the longitudinal and transverse direction, respectively, the support being subjected to the free-field motion as described above. The soil-structure interaction is treated in a quasi-static fashion ignoring inertia effects. The supporting strings, often called moduli of subgrade reaction are given in units of force/displacement per unit length of the tunnel (kN/m/m). Here we use K_a for the longitudinal and K_t for the transverse direction, respectively. Soil-structure interaction effects are then described by means of reduction factors R_N and R_M that are applied to the sectional forces from equations (18)-(20), cf. Kuribayashi *et al.* [10], St. John and Zahrah [14]:

$$N_{\text{max,SSI}} = N_{\text{max}} R_N \tag{21}$$

$$M_{\text{max,SSI}} = M_{\text{max}} R_M \tag{22}$$

with

$$R_N = \frac{K_a}{\frac{E_l A_l}{2} \left(\frac{2\pi}{\lambda}\right)^2 + K_a} \tag{23}$$

$$R_M = \frac{K_t}{E_l I_l \left(\frac{2\pi}{\lambda}\right)^4 + K_t} \tag{24}$$

The shear force is determined via the bending moment:

$$Q_{\text{max,SSI}} = \frac{2\pi}{\lambda} M_{\text{max,SSI}} \tag{25}$$

Since the sectional forces depend on the wavelength λ, we now seek the wavelengths that maximize axial force, bending moment, and shear forces in equations (21)-(25). For simplicity we assume here that the moduli of subgrade reaction are independent of the wavelength λ of the seismic wave. Setting the derivatives of the sectional forces with respect to the wavelength equal to zero we obtain

$$\lambda_N = 2\pi \left(\frac{E_l A_l}{2K_a}\right)^{1/2} \quad ; \quad \lambda_M = 2\pi \left(\frac{E_l I_l}{2K_t}\right)^{1/4} \quad ; \quad \lambda_Q = 2\pi \left(\frac{E_l I_l}{3K_t}\right)^{1/4} \tag{26}$$

with the subscripts N, M, and Q denoting the corresponding sectional forces. Inserting these values into equations (21)-(25) yields the maximum sectional forces:

$$\hat{N}_{\text{max,SSI}} = \frac{1}{4}(2K_a E_l A_l)^{1/2} D_0 \tag{27}$$

$$\hat{M}_{\text{max,SSI}} = \frac{1}{2}(K_t E_l I_l)^{1/2} D_0 \tag{28}$$

$$\hat{Q}_{\text{max,SSI}} = \frac{3}{4}\left(\frac{K_t^3 E_l I_l}{3}\right)^{1/4} D_0 \tag{29}$$

The calculated maximum axial force shall not exceed an upper limit defined by the ultimate soil drag resistance in the axial direction. For given tunnel geometry this value is estimated from the effective shear strength of the soil at the respective depth location and application of an appropriate safety factor.

The above procedure is extremely conservative due to the assumptions made that the ground motion is produced by a single shear wave train, that the wave train impinges upon the structure at the most critical angle, and that the displacement amplitude D_0 is independent of the wavelength while in general ground displacement amplitude decreases with the wavelength. Thus, the equations derived can be considered as upper bounds for the sectional forces.

Often in practice for the sake of simplicity only the axial deformation caused by an angle of incidence $\psi = 45°$ is considered. This is due to the fact that the flexural deformations are small relative to the axial deformations when the wave acts in the critical direction in each case, i.e. angle of wave incidence $\psi = 0°$ for flexural and $\psi = 45°$ for axial deformation.

A point that deserves particular attention refers to the determination of the wavelength dependency of both the moduli of subgrade reaction K_a and K_t, and of the expected displacement amplitude D_0. Even today this is not an easy task.

The moduli of subgrade reaction cannot be evaluated rigorously for a given soil/tunnel system. Some proposals can be found in the literature but show large scatter, the equations depending on the method adopted and the system geometry. St. John and Zahrah [14] derived an expression based on the two-dimensional, plane-strain solution to the Kelvin's problem, i.e. the response of an infinite elastic, homogeneous, and isotropic medium to a static point load. The solution for a sinusoidal load is then determined by an approximate procedure. The resulting equation that applies both for the axial and the transverse horizontal direction reads:

$$K_a = K_t = \frac{16\pi G(1-\nu)}{(3-4\nu)} \frac{d}{\lambda} \tag{30}$$

where G is the shear modulus of the ground. Note that in this expression the subgrade modulus depends on the wavelength, a fact that introduces an additional uncertainty.

Based on a synthesis of earlier proposals by various authors the AFPS/AFTES Guidelines [1] recommend

$$K_a = K_t = G \tag{31}$$

On the other hand, Clough and Penzien [3] recommend

$$K_a = 3G \tag{32}$$

This discrepancy clearly demonstrates the difficulty in assessing realistic values. For critical projects it is recommended to compute the response of the actual soil-structure system by means of a finite-element analysis that automatically yields different values for both the axial and the transverse springs.

An alternative method is presented next. It is based on a quasi-static analysis and considers in an approximate manner the relevant parameters, i.e. geometry of the foundation, embedment depth, and non-linear effects of soil-structure interaction. It is applicable to tunnels in shallow depths, e.g. immersed tubes crossing shallow waters. We consider a single tunnel element with a given aspect ratio in the order of 5:1 to 10:1. Interaction between the tunnel elements is neglected. Perfect bonding between plate (tunnel) and foundation soil, and linear-elastic theory are assumed. The shear modulus of the foundation soil is taken equal to the one determined from the equivalent nonlinear free-field seismic site response analysis of the soil deposit, i.e. it is a reduced shear modulus including the effects of cyclic loading. With these values the modulus of subgrade reaction is determined first for a surface foundation. The effects of tunnel embedment are then taken into account leading to an increase of the values of subgrade reaction. An additional increase is associated with the assumption made that the shear modulus degradation of the soil underneath the tunnel will be smaller than under free-field conditions due to the effects of the tunnel-soil interaction.

The equation for the spring constant per unit tunnel length reads:

$$K_m = \alpha_m \bar{G}_m \tag{33}$$

where m refers to the vibration mode, i.e. longitudinal, transverse horizontal, and vertical. $\bar{G}_m$ is the mean effective shear modulus of the ground in the vicinity of the tunnel element, depending also on the vibration mode.

The factor α_m is composed of three terms:

$$\alpha_m = \alpha_{0,m}\alpha_{1,m}\alpha_{2,m} \tag{34}$$

where $\alpha_{0,m}$ represents the response of the surface foundation, the factor $\alpha_{1,m}$ reflects the effect of embedment, and $\alpha_{2,m}$ the effect of reduced seismic strain amplitude of the near surface ground due to the presence of the rigid tunnel structure. All three factors depend on the depth variation of the effective strain compatible shear modulus, on the geometry of the foundation, and on the vibration mode.

The formulas given by Gazetas [4] are used to determine numerical values for the static stiffness of the rectangular plate, i.e. $\alpha_{0,m}$ and $\alpha_{1,m}$. These formulas have been derived by approximating rigorous solutions of the soil-structure interaction problem, i.e. they capture all essential features. For the calculation of the embedment factor the actual sidewall-soil contact surface is used, which consists only of the two sidewalls of the tunnel elements. For horizontal motions the parameter α_2 is roughly in the order of 1.1 to 1.2, e.g. Kuesel [9] reports that analyses for San Francisco Trans-Bay Tube yielded a 15% reduction of the free-field deformation due to the tunnel rigidity. For vertical motions α_2 will be higher due to the lower level of seismic excitation and accordingly the lower shear modulus degradation. A reasonable estimate would be an increase of 20% with respect to the value for horizontal motions.

The region of soil underneath the foundation which participates in the tunnel-soil interaction, and consequently the value of the mean effective shear modulus $\bar{G}$ in that region, depends on the vibration mode: Vertical vibration modes reach deeper regions than horizontal ones. Assuming an increase of soil stiffness with depth, $\bar{G}$ for the vertical mode will be larger than for the horizontal modes. The corresponding values can be calculated by means of a static deformation analysis of the foundation. The actual depth-distribution of soil stiffness is taken equal to the strain compatible shear modulus profile as determined from the non-linear seismic site response analysis. The calculated values for the subgrade moduli shall always be regarded as best estimates, and a variation of at least $\pm 30\%$ is suggested for the structural design. Note that in the above approximate methodology the subgrade modulus is independent of the wave length.

6.3   Influence of transverse joints

The above derivations assume a continuous tunnel lining. Transverse joints are placed between tunnel elements to reduce seismic strains. These joints can be either open or flexible. The latter are typically found in submerged tunnels connecting tunnel elements and are of various types. We first consider *flexible* joints. For the analysis we adopt the simplified numerical model introduced by Hamada *et al.* [6] as depicted in Figure 3. We assume that the axial strain of the tunnel due to push-pull deformation dominates the tunnel behavior compared to the axial strain component due to bending around the vertical axis. We further assume that the ground normal strain ε_g is uniformly distributed along the tunnel axis. Then the dynamic strains in the tunnel ε_t and the associated relative displacement of the flexible joint δ_j are given by:

$$\varepsilon_t(x) \;=\; \left(1 - \frac{\cosh \beta x}{\cosh \beta L_j}\left(1 - \frac{\tanh\left(\beta L_j/2\right)}{\left(\beta E_l A_l/2 k_j\right) + \tanh\left(\beta L_j/2\right)}\right)\right)\varepsilon_g \quad (35)$$

$$\frac{\delta_j}{L_j} \;=\; \frac{\tanh\left(\beta L_j/2\right)}{\left(\beta L_j/2\right) + \left(k_j L_j/E_l A_l\right)\tanh\left(\beta L_j/2\right)}\varepsilon_g \quad (36)$$

with

$$\beta = \sqrt{K_a/E_l A_l} \quad (37)$$

In the above expressions L_j is the distance between two adjacent joints, i.e. the tunnel element length, K_a is the subgrade modulus in the axial direction, and k_j the axial spring stiffness of the flexible joint. The maximum strain value is obtained at the midway point with $x = 0$. The limiting case of an *open* joint can be simulated by setting $k_j = 0$.

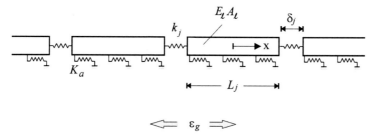

Figure 3: Simplified model for longitudinal seismic response, Hamada *et al.* [6]

The presence of open joints will cause joint separation. The associated gap displacements during an earthquake may produce undesirable effects such as leakage of water-tight seals. Joint details shall be selected accordingly to accommodate the expected displacements.

6.4   Global tunnel response

The axial and flexural deformations can of course be determined numerically on a global tunnel model consisting of a beam supported by springs, that reflect the soil structure interaction effects, and subjected to a spatially varying free-field soil deformation pattern. Ignoring ground motion incoherence effects, the spatial variability of the ground motion can be approximately simulated by the passage of shear waves without the presence of wave scattering. The angle of incidence is usually set equal to 45° with respect to the longitudinal tunnel axis. The wave passage takes place with an apparent velocity c. The seismic motion is imposed in terms of the free-field displacements on the soil ends of the springs with a phase shift equal to the apparent wave-passage velocity, cf. Figure 4. Inertia effects of the beam are neglected. It can be shown that axial force and bending moments on the tunnel cross-section increase with decreasing apparent wave velocity c. Typically, apparent wave passage velocities range between 1000 m/s and 2500 m/s, cf. Hamada *et al.* [6].

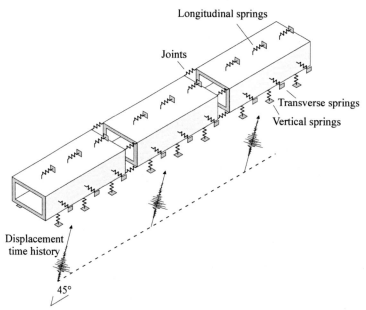

Figure 4: Model for global tunnel response

7   Ovaling and racking deformation analysis

Ovaling of mined circular or racking of rectangular cut-and-cover tunnel linings represents the most critical deformation pattern, Wang [15]. Since the size of a typical cross-section is small compared to the dominant wavelengths of the seismic motion, it can be assumed that the induced shear strain is constant over the height of the tunnel structure. Further it is assumed that inertia effects in soilstructure interaction are negligible, thus allowing a quasi-static analysis.

First, the expected maximum free-field shear strain $\gamma_{\max}$ due to vertically propagating shear waves is determined. For deep and relatively uniform ground an estimate is obtained by the simplified Newmark method:

$$\gamma_{\max} = \frac{V_{s0}}{\bar{c}_S} \tag{38}$$

where V_{s0} is the peak particle shear wave velocity and $\bar{c}_S$ is the effective shear wave propagation velocity in the depth range of the structure accounting also for the effects of shear strain amplitude. In the preliminary design stage estimates for $\bar{c}_S$ can be obtained, for example, by applying to the small-strain values c_S an appropriate reduction factor in dependence of the maximum earthquake acceleration, as described for example by Eurocode EC8, Part 5.

The shear strain at the elevation of the tunnel axis multiplied by the height of the structure yields an approximation for the difference of the ground displacement between the top and bottom elevation of the structure:

$$\Delta_{ff} = \gamma_{\max} h \tag{39}$$

where h is the height (or diameter) of the tunnel structure.

In the final design a numerical seismic site response analysis for vertically propagating shear waves shall be conducted, e.g. using the code SHAKE or similar. The difference of maximum shear strains at the top and bottom elevations of the structure is determined directly from the analysis, cf. Figure 5. The differential displacement is calculated analogously.

We next consider a circular lining of diameter d. Due to the imposed free-field strain it will rack into an oval shape yielding a change of diameter Δ in the diagonal principal directions (45°), cf. Figure 1. Two limiting cases are investigated: i) lining distortion stiffness is almost equal to that of the surrounding ground, and ii) lining stiffness is small compared to that of the ground. To model the first case we determine the diametric strain of an imaginary circle in a non-perforated medium:

$$\frac{\Delta_{ff}}{d} = \pm\frac{\gamma_{\max}}{2} \tag{40}$$

The second case is modeled by a cylindrical cavity yielding a much larger value for the diametric strain

$$\frac{\Delta_c}{d} = \pm 2\gamma_{\max}(1 - \nu) \tag{41}$$

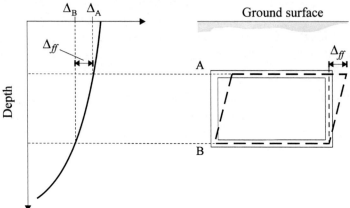

Figure 5: Free-field racking deformation on a rectangular frame

For a typical value $\nu = 0.25$ the distortion from equation (41) is 3 times higher than that from equation (40). Hence, the relative stiffness between the lining and the medium governs the response. A solution that considers the soil-structure interaction effect is presented by Penzien [12]. Slippage is assumed not to occur. A lining racking ratio R_r defined as the ratio of the diameter change of the lining Δ_l to the respective value under freefield conditions Δ_{ff}

$$R_r = \frac{\Delta_l}{\Delta_{ff}} \tag{42}$$

is derived:

$$R_r = \frac{4(1-\nu)}{1 + k_l(3 - 4\nu)/k_s} \tag{43}$$

where k_s is the generalized stiffness of the soil displaced by the lining, and k_l is the generalized stiffness of the lining. These stiffnesses can be evaluated using the relations

$$k_s = \frac{2\bar{G}}{d} \tag{44}$$

$$k_l = \frac{48 E_l I_l}{d^4 \left(1 - \nu_l^2\right)} \tag{45}$$

where $\bar{G}$ is the strain-compatible soil shear modulus that is determined from the strain-compatible shear wave velocity using equation (1). Note that the racking ratio equals unity when the lining stiffness equals the soil stiffness, while as k_l approaches zero (cavity) the racking ratio R_r approaches $4(1-\nu)$.

The racking response of a rectangular lining of height h and width b can be derived analogously, whereby Δ_l is the relative displacement between top and bottom of the lining, cf. Figure 1. Equation (40) and (41) may be used also for this case. However, the stiffnesses are different, Penzien [12]. The stiffness of the soil displaced by the lining is now given by

$$k_s = \frac{\bar{G}}{h} \tag{46}$$

The lining lateral stiffness k_l is obtained from simple static analysis of the box structure without the surrounding soil under simple boundary conditions as shown in Figure 6, whereby the stiffness coefficient k_l is that shear stress τ_l which will produce the corresponding unit racking displacement of the lining under plane strain conditions.

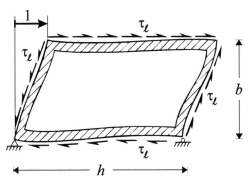

Figure 6: Simple model to calculate lateral lining stiffness, Penzien [12]

The last step consists in applying the racking displacement as input in the structural model of the tunnel to calculate the shear, moments and other structural design parameters. The racking displacement is simulated by means of either i) a concentrated force at the roof level or ii) a triangular pressure distribution along the walls of the tunnel. The former is more appropriate for deep tunnels and the latter for shallow ones, but it is recommended to examine both cases in the analysis, Wang [15].

The boundary effect due to the vicinity to the ground surface is considered negligible for burial depths (ground surface to mid-height of lining) greater than 1.5 times the gross height of the structure. A reduction of the burial depth to half the tunnel height yields approximately a 20% increase of the racking ratio, cf. Wang [15].

Alternatively, a 2D dynamic finite element analysis may be carried out using appropriate codes such as FLUSH. Like in SHAKE, the nonlinear soil behaviour is represented by a visco-elastic equivalent model described by strain-dependent shear modulus and damping ratio. The structure is modelled by elastic beam elements with frequency independent damping properties. Laterally, appropriate transmitting boundaries allow for the outward propagation of waves. The earthquake motion is applied in the base of the model. This type of transverse analysis shall always be conducted for the final design of large shallow tunnel structures with irregular geometry such as cut-and-cover structures sheltering urban motorways or train lines.

References

[1] AFPS/AFTES: Guidelines on Earthquake Design and Protection of Underground Structures, 2001.

[2] Arango, I, Kulesza, R. and Wu, C.L.: San Francisco's MUNI metro underground extension, Proc. 11th European Conference on Earthquake Engineering, 1998.

[3] Clough, R.W. and Penzien, J.: Dynamics of Structures, 2nd Edition, McGraw-Hill, 1993.

[4] Gazetas, G.: Formulas and charts for impedances of surface and embedded foundations. *J. Geotech. Engrg.*, ASCE, Vol. 117, pp. 1363-1381, 1991.

[5] Dowding, C. H. and Rozen, A.: Damage to rock tunnels from earthquake shaking, *J. Geotech. Engng Div.*, ASCE, Vol. 104, No. GT2, pp. 175-191, 1978.

[6] Hamada, M., Shiba, Y. and Ishida, O: Earthquake observation on two submerged tunnels at Tokyo Port, Soil Dynamics and Earthquake Engineering Conference, Southampton, pp. 723- 735, 1982.

[7] Hashash, Y.M.A., Hook, J.J., Schmidt, B., and Yao, J.I.C.: Seismic analysis and design of underground structures, *Tunnelling and Underground Space Technology*, Vol. 16, pp. 247-293, 2001.

[8] Kramer, S.L.: Geotechnical Earthquake Engineering, Prentice-Hall, 1996.

[9] Kuesel, T. R.: Earthquake design criteria for subways, *J. Struct. Div.*, ASCE, Vol. 95, No. ST6, pp. 1213-1231, 1969.

[10] Kuribayashi, E., Iwasaki, T., and Kawashima, K.: Dynamic behaviour of a subsurface tubular structure, *Bulletin New Zealand Soc. for Earthq. Eng.*, Vol. 7, No. 4, 200-209, 1974.

[11] Newmark, N.M.: Problems in wave propagation in soil and rock. Proc. Int. Symp. Wave Propgation and Dynamic Properties of Earth Materials, New Mexico, Univ. of New Mexico Press, 1967.

[12] Penzien, J.: Seismically induced racking of tunnel linings, *Earthquake Engng. Struct. Dyn.*, Vol. 29, pp. 683-691, 2000

[13] Power, M.S., Rosidi, D. and Kaneshiro, J.Y.: Seismic vulnerability of tunnels and underground structures revisited. North American Tunneling '98, (Ozdemir, Ed.), Balkema, Rotterdam, pp. 243-250, 1998.

[14] St. John, C. M. and Zahrah, T. F.: Aseismic design of underground structures, *Tunnelling and Underground Space Technology*, Vol. 2, No. 2, pp. 165-197, 1987.

[15] Wang, J.-N.: Seismic Design of Tunnels: A State-of-the-Art Approach, Monograph 7., Parsons, Brinckerhoff, Quade and Douglas Inc., New York, 1993.

[16] Wells, D.L. and Coppersmith, K.J.: New empirical correlations among magnitude, rupture length, rupture width, rupture area and surface displacement, *Bull. Seism. Soc. Am.*, Vol. 84, No. 4, pp. 974-1002, 1994.